# ANAN
## Stream of Living Water

Bonnie Demerjian

Bonnie Demerjian
Color photographs by Ivan Simonek

Stikine River Books
Wrangell, Alaska

Published by
Stikine River Books
P.O. Box 1762
Wrangell, Alaska 99929 USA




ISBN 10: 0-9776792-1-7
ISBN 13: 978-0-9776792-1-8
Library of Congress Control Number: 2007902338

Cover and book design by Matt Knutson, InterDesign
Interior color photographs by Ivan Simonek, Mark Emery and Bonnie Demerjian
Maps and drawings by Peter Branson

Cover photograph by Ivan Simonek. Black bear bathing in Anan Creek.

Printed by Everbest Printing Co., Ltd, Nansha, China

# Contents

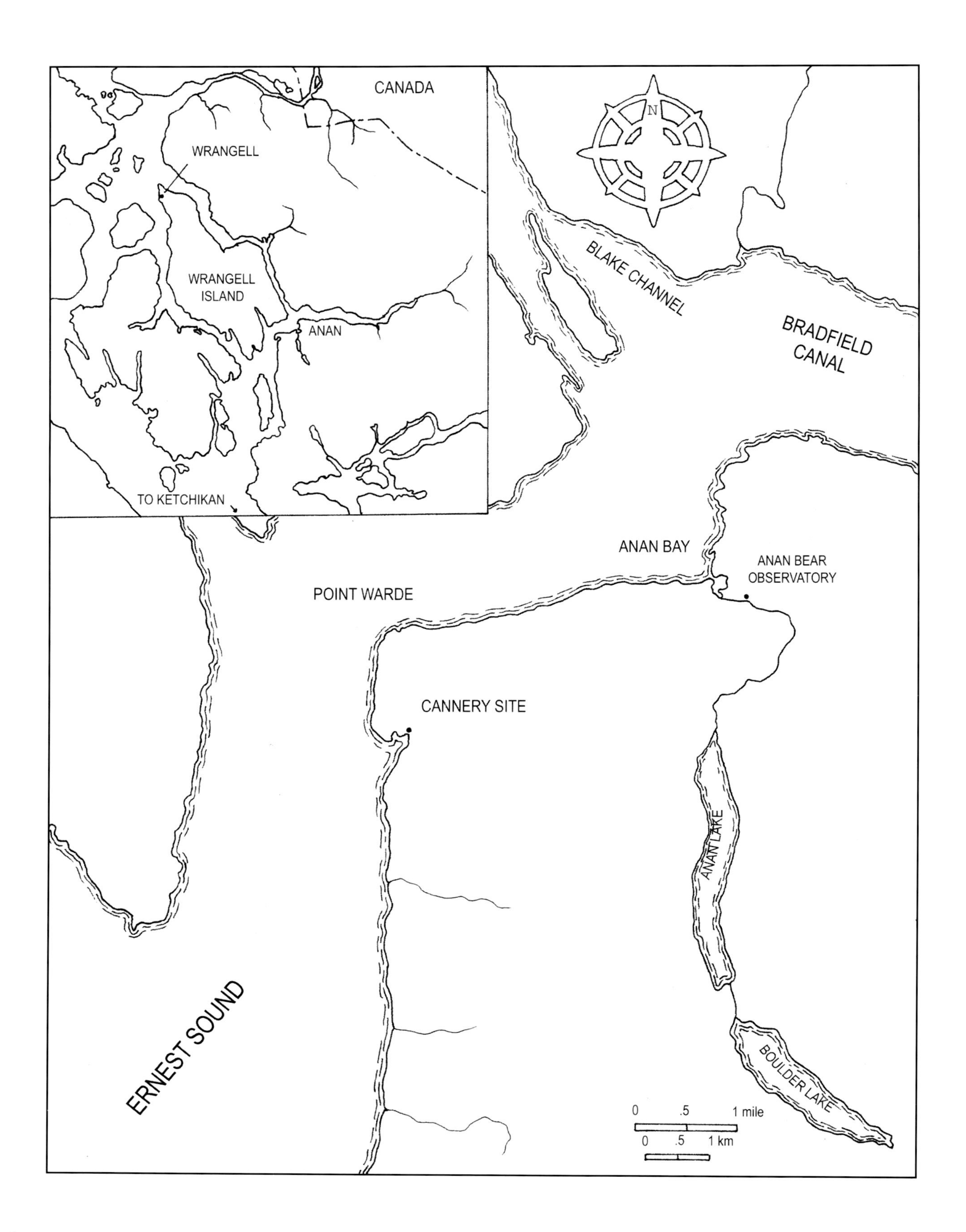

CANADA
WRANGELL
WRANGELL ISLAND
ANAN
TO KETCHIKAN
N
BLAKE CHANNEL
BRADFIELD CANAL
ANAN BAY
ANAN BEAR OBSERVATORY
POINT WARDE
CANNERY SITE
ANAN LAKE
BOULDER LAKE
ERNEST SOUND
0 .5 1 mile
0 .5 1 km

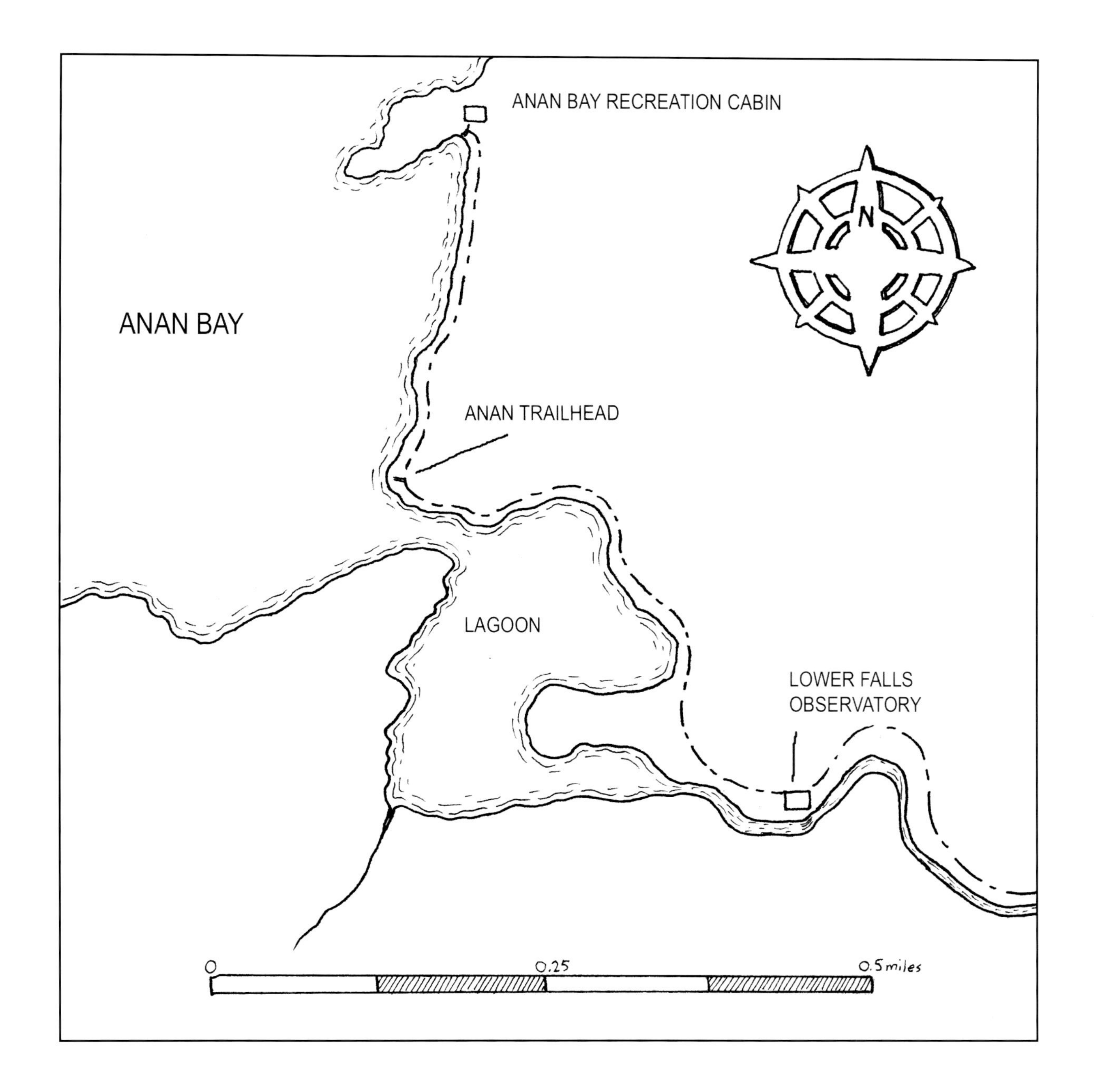
ANAN BAY RECREATION CABIN
N
ANAN BAY
ANAN TRAILHEAD
LAGOON
LOWER FALLS
OBSERVATORY
0
0.25
0.5 miles

✦ CHAPTER ONE

# Here Earth and Water Seem to Strive Again*

What could be simpler than a wooded stream rippling through an unassuming valley? True, it attracts bears, brown and black, with an abundance of salmon. Other creatures, too, gravitate to its waters and banks to feast on the leavings of bears, decayed carcasses of spawned-out fish. Humans are among those attracted to the stream, Native people from past millennia and today's visitors. This is an modest creek, Anan, not so different from hundreds or even thousands of others in Southeast Alaska, Yet, by walking its trail and studying its kaleidoscope of land and water, its marine, plant, animal and human life, we can begin to glimpse extraordinary interdependence in that singular ecosystem, the salmon stream. Anan Creek is an abundant and complex place well worth knowing.

The creek is located on the northern shore of the Cleveland Peninsula, an arm of the Southeast Alaska's mainland, positioned south of Wrangell and north of Ketchikan. Anan's story begins with a clash of titans as tectonic movement led the North American and Pacific plates, slowly but inexorably over 500 million years, in a grinding, elephantine dance. Blocks of the earth's crust, some originating at the equator or further, moved northward and east. Some of this well-traveled material was subducted or forced beneath the edge of the North American plate, melting existing rock and raising mountains as it sank while accreting or welding onto the North American plate. Other blocks kept moving north.

Bonnie Demerjian

*Anan Creek in early autumn.*

Geologists call these successive waves of new material, unrelated to the surrounding material, terranes. They form southeast-north-

* *Alexander Pope, "Windsor Forest"*

west trending rock belts the length of Southeast Alaska. Ice from the Cordilleran Ice Sheet also shaped this land in successive flows, spilling from the crests of the Coast Range, incising valleys and marine trenches. In southern Southeast Alaska the most recent cold period, the Wisconsin glaciation, reached its maximum between 21,000 and 13,000 BP (Before Present). By 13,000 BP much of the region was clear of ice, leaving the drowned valleys (fiords) and uncounted islands of today's Alexander Archipelago. As a result of heavy burdens of ice followed by deglaciation, the coastline of Southeast Alaska has changed considerably during the last 30,000 years as both sea levels and land rose and fell in a seesaw movement that continues in some places even today.

As the ice retreated plants began to colonize the newly bared land. Study of pollen cores from Southeast lakes and peat deposits is bringing to light the region's early environmental history. Thus far, it appears that following deglaciation in the southern part of the region, tundra and open pine forest appeared, perhaps as early as 12,500 BP, followed by alders and Sitka spruce. In other areas, a tundra of sedges, grasses and herbs was succeeded by dwarf willows and heath, then trees in approximately this order: alder, lodgepole pine, Sitka spruce, western hemlock and mountain hemlock.

Ivan Simonek

*This view of the Stikine Icefield today gives a glimpse of how much of Southeast Alaska would have looked during the most recent ice age*

Traditionally, species of both plants and animals recolonizing Southeast Alaska were assumed to have originated either to the north or south of Alaska. Berengia, the wide land bridge from Asia, was the northern source while the other was present-day Washington and points south of the ice sheet. Ongoing research into Southeast Alaska's glacial history, however, is uncovering a number of probable ice-free zones or refugia nearer to home where the hardiest species persisted during glacial times. Among these sites are areas of the inner continental shelf, parts of the westernmost islands of the archipelago, unglaciated ocean-facing slopes and nunataks or isolated mountain peaks on several of the region's largest islands. Another nearby refugium may have been shallow Hecate Strait between the Canadian mainland and the Queen Charlotte Islands. A fossil record of animal species dating as early as 50,000 BP is emerging from caves on and near Prince of Wales Island. Brown bear, ermine, chum salmon, shore pine and subapline pine are among the species believed to be relicts or survivors of the Wisconsin glaciation. Humans, also, used these caves at least ten thousand years ago. Local refugia and those to the north and south enabled rapid recolonization of animals and plants into the region. Other species arrived later from the mainland interior once valley corridors were cleared of ice.

Fish repopulated Southeast Alaska from Berengian river systems in the north and the Columbia River to the south. Today, the Stikine River near Wrangell acts as a rough dividing line between the Berengian populations of fishes and the Columbian. Inland freshwater routes, some of whose courses have changed considerably from early postglacial days, also provided paths for fish recolonization.

To better understand the wide variety of landforms in sprawling and geologically complex Southeast Alaska, the

U.S. Forest Service has divided the region into ecological subsections that help define the landforms, rocks and soils. The majority of the Anan Creek drainage is in the Bell Island Granitics subsection characterized by long, narrow bays echoing the northwest-southeast orientation of glacial movement. Anan and Boulder lakes and several smaller unnamed lakes drain into salt water via the two narrow forks of Anan Creek. These are remnants of glacial retreat. Rounded hills and mountains up to 3,000 feet also are evidence of glacial scouring in this subsection. East of Anan Creek further up the Bradfield Canal, the land changes to rugged sedimentary and volcanic formations and numerous short, steep streams - characteristics of the Eastern Passage Complex. Anan Creek is one of the largest streams in the region. It is twelve and a half miles long and drains forty-eight square miles from heights of 1600 feet.

Over rock and through forest, percolating down through soils both thin and densely organic flows water, Southeast Alaska's most constant element. The moist maritime climate of the region nourishes lush forests and wetlands. High latitude makes the region a temperate rainforest, uniquely productive and globally important. This temperate rainforest is an ecosystem, a community of organisms and their physical environment interacting as a unit. Water is a defining feature of this region's ecology and salmon – that quintessential emblem of the Pacific Northwest – is an important actor in that watery arena. Salmon, has, in fact, been called the keystone species of the region. A keystone species is one that has a disproportionate effect on its environment relative to its abundance. Such an organism plays a role in its ecosystem that can be compared to the role of a keystone in an arch — if the keystone in the arch is removed the arch collapses. Similarly, an ecosystem may experience a dramatic shift if a keystone species is removed, even though that species may be only small part of the ecosystem in terms of its numbers.

## Discovering and Naming Anan Bay

Captain George Vancouver first reported Anan Bay in 1793 as part of his extensive explorations of Southeast Alaska. He was charged by the British with examining every inlet leading east or north from California to Alaska and determining once and for all the existence of a Northwest Passage. In his ships *Discovery*, *Chatham* and several smaller vessels he and his crew tediously but thoroughly explored and charted the coast. As they went he named 338 geographical features of the region, often for his officers and crew as well as for royalty and officers of the Admiralty. In central Southeast Alaska Vancouver named a number of places after persons who had served with the British army in Ireland.

Under Vancouver's command Lieutenant James Johnstone led a scouting party that discovered and explored the Bradfield Canal in one of the launches, noting that they were "much impeded by adverse winds, and very unpleasant weather." A canoe with three Natives accompanied the men for part of their survey up the Canal and then left after being given "some trivial presents." Nearby Point Warde was named by Vancouver for General George Warde, Commander-in-chief in Ireland from 1791 to 1793 who hailed from Bradfield, Berkshire. Inconspicuous Anan was not remarked upon at this time. Perhaps the explorers never realized the fishy wealth they passed by. Perhaps they didn't care.

Lt. Commander Albert Sydney Snow in charge of the United States Coastal and Geological Survey steamer, *Patterson*, surveyed Southeast Alaska in 1886. Under his command the crew took soundings and mapped the Bradfield Canal. While doing so, he applied the name Anan to a small bay east of today's Anan Bay (the unnamed bay into which flows Canal Creek), using its Tlingit designation, which meant "Sit Down Town," perhaps referring to the several clans that camped nearby in summer and fall.

In 1904 the present Anan Bay received the name Humpback Bay, an obvious choice, from H.C. Fassett of the United States Bureau of Fisheries. This term was retained in the Coast Pilot through the 1960s where it states: "Humpback Bay and Anan Bay are two open bights on the south shore of the Bradfield Canal. A large salmon stream empties into Humpback Bay." It also noted that the Forest Service maintains a bear observatory on Anan Creek where many people come to view black bear feeding on salmon. No mention is made of brown bear. Although Humpback Bay is occasionally found on older charts, Anan Bay has now resumed its traditional designation.

*The rugged peaks of eastern Bradfield Canal, characteristic of Eastern Passage Complex geology.*

Ivan Simonek

Ivan Simonek

*Anan's lower falls.*

The lives and deaths of salmon link land and water, feeding flora and fauna alike while returning nutrients to both the soil and water. Salmon are anadromous fish, spawning in freshwater streams and lakes after reaching maturity in salt water. Young salmon spend anywhere from a few months to a few years in fresh water, depending on the species and other factors, before heading downstream to the estuaries and ocean. However, over ninety-five percent of a salmon's mass accumulates in the ocean. This unique life history results in a massive flow of energy, termed a "conveyor belt" by aquatic ecologist David Schindler, from the ocean to freshwater where the salmon's nutrient and energy-rich carcasses are deposited. This new understanding reverses the traditional view that nutrients only flow downstream.

Scientists have used nitrogen-sampling techniques to determine how much nitrogen in an ecosystem comes from salmon by measuring stable isotopes. Nitrogen from the land and freshwater is primarily the lighter $^{N}14$ while that from salt water is the heavier $^{N}15$. By comparing the amounts of each in any material from salmon to alder leaf, scientists can discover where salmon-borne nitrogen is amassed. Carbon, sulfur and phosphorus are also returned upstream by salmon.

Ivan Simonek

*Bears are the most charismatic of salmon lovers but are only one of many species that rely on that fish for sustenance.*

How do these nutrients move "uphill" from the ocean? A black bear fishing beneath Anan's observatory snatches a returning pink salmon from the turbulent waters. While she may consume a part of that on the spot, she often will drag a partially consumed carcass up into the woods to finish later or offer to her cubs. When they have eaten their fill scavengers such as mink, marten, eagles, gulls, mergansers and beetles will consume the remains. As the remnants decompose nutrients leach into the soil to be taken up by streamside vegetation. Sitka spruce grow faster and densities of songbirds may prove to be greater near salmon-bearing streams. The presence of salmon also may influence bear population and health. Brown bear numbers, for example, are eighty times denser and their body size larger in coastal areas with salmon streams than those in the interior. In fact, studies indicate that the presence of salmon increases the total productivity of a stream and they are, perhaps, essential to the entire watershed. Scientists have found that in Southeast Alaska at least sixty species of wildlife besides bears use salmon for food as they feast on eggs, carcasses and living fish. Another study in Washington found that if microbes and invertebrates are included, salmon provide food for at least 137 species.

Recycling energy and matter is now understood to be a key principal of ecology. As Fritjof Capra, physicist and systems theorist, states, "All organisms in an ecosystem produce wastes, but what is waste for one species is food for another, so that wastes are continually recycled and the ecosystem as a whole generally remains without waste." (Capra, p. 177)

Some species even coordinate life events with that of the spawning salmon. Mink and river otters, for example, time their reproductive cycles to that of salmon returns, undergoing lactation when nutritious salmon carcasses are most available. Juvenile eagles, a common sight in the trees of Anan Creek and lagoon, are just beginning to forage for themselves when the equivalent of a fast food restaurant arrives in the form of spawning salmon. Juvenile coho and Chinook salmon and rainbow trout all consume either salmon eggs or other juveniles. Evidence suggests that higher levels of the same omega-3 fatty acids that contribute to human health may also increase salmon growth rate and size.

Shifts of just one or two degrees in climate can affect the nutrients taken in by salmon and cause changes in population size. These changes are reflected in nutrient levels found in core samples taken from Alaskan lakes where sockeye salmon spawn and die. Here, high levels of plankton coincide with high numbers of salmon. Aquatic ecologist Bruce Finney of the University of Alaska, looking beyond human influence, has found that salmon populations in the North Pacific dropped sharply two thousand years ago. They stayed low for about eight hundred years, increasing gradually and peaking around 1900. That date marked, perhaps not coincidentally, the beginning of intensive salmon fishing in Alaska. Other studies of salmon populations in the Pacific Northwest have shown an increase in Chinook salmon returns during cool, wet summers and a decrease during warm, dry ones. Fisheries managers formerly considered environment to be constant and consequently harvest limits were unvaried. Today we know differently and science is scrambling to understand fish and their responses to climate.

In the late 1940s there were calls from Alaskan politicians, spurred by fishery managers and fishermen, to cull bears in order to reduce the perceived economic damage that they might be wreaking on salmon populations. They were following a familiar pattern in which a single "valued" species is emphasized to the neglect or extermination of others. Fortunately for the bears and the entire ecosystem, the enthusiasm for bear control never came to pass. Bald eagles, Dolly Varden trout, seals and sea lions, too, were looked upon as hazardous to salmon numbers. Between 1917 and 1953 a bounty of 50 cents a foot was offered for eagle talons with an estimated 100,000 to128,000 Alaskan birds recorded in bounty books. By contrast, researchers are now engaged in study of the interdependence of all of a salmon stream's elements—flora, fauna, soil and water. This meshing of life means that streams in the Pacific Northwest that have lost their salmon populations are seeing changes that reach far beyond waters empty of fish. In its heyday the Columbia River is estimated to have hosted 500 million pounds of salmon. These contributed hundreds of thousands of pounds of nitrogen and phosphorus to the river's drainage. The Columbia River Valley is now starved for lack of nutrients by dams, irrigation and habitat loss and its salmon run is history. It appears, then, that the ecological balance achieved by a salmon stream over millennia is critical for the health of all the stream's inhabitants and the very land itself.

Besides being unwitting chemists, salmon are also engineers. Rather than passively reacting to their stream environment, salmon modify it, triggering several changes in the streambed. To prepare a site for depositing their eggs, female salmon excavate nests called redds. With their tails they furiously churn up silt that would otherwise smother the eggs, sweeping it aside to create a clean gravel bowl about twenty-five inches square in the case of pink salmon. This tilling activity, particularly in a stream as dense with salmon as Anan, can change the shape and composition of the streambed. The ridges and depressions of the redds change water flow patterns and over time may cause the channel to migrate. Removing the fine sediments may also increase the stream's speed and oxygen flow, promoting higher egg survival.

Recent research activity is peeling back layers of mystery that veil the ecology of a salmon stream but much remains to be understood. What we have previously identified as individual components is now revealed to be a web of relationships. We are slowly, sometimes painfully, learning that we must care for everything because each part is vital to the health of the whole.

# Plant and animal species at Anan

Here is a partial list of plant and animal species in the Anan watershed. These are typical species of Southeast Alaska's moist, low-lying coastal areas. Many of these species at the Creek or in nearby waters rely on salmon eggs, alevin, fry, adults and carcasses of spawned-out adults:

**Plants**

licorice fern
oak fern
narrow beech fern
deer fern
lady fern
shield fern
maidenhair fern
Braun's holly fern
skunk cabbage
deer heart lily (false lily of the valley)
watermelon berry (clasping twisted-stalk)
broadleaved (two-flowered) marsh marigold
yellow marsh marigold
western buttercup
little buttercup
trifoliate foam flower
five leaf bramble
goat's beard
large avens
enchanter's nightshade
beach lovage
Indian celery
bunchberry
cleavers

Ivan Simonek

*This great blue heron also feeds on salmon carcasses.*

Bonnie Demerjian

*A brilliant Oregon crabapple brightens Anan lagoon in fall.*

yarrow
arrow-leaved groundsel
Western rattlesnake root
salmonberry
thimbleberry
oval-leaf blueberry
early (Alaskan) blueberry
red huckleberry
rusty menziesia
devil's club
trailing black currant
stink currant
Pacific red elderberry
western red-cedar
yellow-cedar
Sitka spruce
western hemlock
mountain hemlock
red alder
Oregon crabapple
red belt fungus
sulfur shelf (chicken of the woods)
common gel bird's nest (fairy cups)
witch's butter (yellow jelly)

**Fish**
Chinook (king) salmon
chum (dog) salmon
pink (humpback) salmon
sockeye (red) salmon
cutthroat trout
steelhead trout
Dolly Varden trout

**Birds**
American dipper
bald eagle
Barrow's goldeneye
belted kingfisher
Bonaparte's gull
chestnut-backed chickadee
common loon
common merganser
common raven
dark-eyed junco
glaucous-winged gull
golden-crowned kinglet
great blue heron
hairy woodpecker
harlequin duck
hermit thrush
herring gull
marbled murrelet
northwestern crow
red-breasted sapsucker
ruby-crowned kinglet
song sparrow
Steller's jay
Swainson's thrush
varied thrush

**Mammals**
black bear
brown bear
harbor seal
mink
red squirrel
river otter
Sitka black-tailed deer

Ivan Simonek

*Eagles by the score descend on Anan Creek to scavenge salmon carcasses.*

Ivan Simonek

*A brown bear sow perches on a rock to spy pink salmon as they swim upstream.*

✦ CHAPTER TWO

# A Pretty Kettle of Fish*

The pink salmon, also called humpback or humpy from the prominent hump that forms on the back of a spawning adult male, has been called the "bread and butter" fish of the Alaskan salmon industry. It is the most abundant salmon and is found in approximately three thousand pink salmon streams in Southeast Alaska. Of those streams, Anan Creek has been recognized since prehistory as having one of the major pink salmon runs in the region. It is, in fact, one of the world's largest pink salmon producing systems. Native peoples for several millennia and commercial fishermen since the nineteenth century have relied on the annual, predictable harvest of this most plentiful of Pacific salmon. Anecdotes abound such as this from 1949: that memorable year a record number of pinks attempted to scale the lower cascade at Anan. One old timer remembers that those failing the leap would fall backwards, not into water but onto other salmon, skidding along their backs until finding open water again. Records from a weir at Anan, constructed to count returning salmon, show escapements exceeding 600,000 fish in certain years. "Escapement" is a fishery management term referring to the portion of an anadromous fish population that is permitted to escape commercial and recreational fisheries and reach the freshwater spawning grounds. Harvest levels from early in the century were estimated at two to three million fish from Anan Creek. In 1972 more than a million and a half pinks bound for that stream were harvested in the nearby Bradfield Canal/Ernest Sound/Clarence Strait area. Those robust numbers continue today.

Ivan Simonek

*A brown sow and cubs forage at their favorite fishing hole at the mouth of Anan Creek.*

Pink salmon occur widely in streams and rivers north of 40 degrees N. latitude from Asia around the Pacific Rim to North America. Georg Wilhelm Steller, the German naturalist who accompanied the Russian Vitus Bering on his voyage of North American discovery in 1741- 42, was the first European to list the species of Pacific salmon. The pink salmon's scien-

* *Gilbert and Sullivan,* Iolanthe

tific name, *Oncorhynchus gorbuscha*, reflects the pink salmon's Asiatic lineage and Steller's Russian link since gorbuscha means humpback in Russian.

Scientists are divided on whether this fish evolved from an ancestral freshwater form or was of pelagic (ocean dwelling) origin. The oldest fossil salmon dates from about fifty million years ago but the imagination boggles at its descendant, the sabertooth salmon. This ten-foot creature weighed five hundred pounds and sported in waters five to six million years ago. Modern salmon evolved about two million years ago, even then preferring chill rivers and oceans.

In Southeast Alaska, pinks are divided into two groups, northern and southern, with Sumner Strait being the approximate boundary between the two. Pinks spawning in the north enter the archipelago as they return to their home stream from the ocean through Icy Strait or northern Chatham Strait while the southern stock arrives via Dixon Entrance or Sumner Strait.

Pinks spawn in the lower reaches of streams and, in some areas, the intertidal zone. Although they occur in large river systems, they principally inhabit short, fast flowing coastal streams. These salmon rely on fresh water less during their life history than all other species of salmon (Chinook/king, sockeye/red, coho/silver and chum/dog and two Asian species not found in Alaska). They begin migrating toward the ocean just a few hours after emerging from their natal gravel and spend less than a week in fresh water.

Elliott Fisher, courtesy of Ketchikan Museum, Tongass Historical Society Collection.

*A weir across Anan Creek during the 1930s and 40s funneled returning salmon through a narrow opening. Above it sits a salmon counter from the U.S. Bureau of Fisheries. The counter is holding a "tallywacker" or clicker to keep track of the number of fish passing up through the weir.*

Unique to pink salmon is their two-year life span. They return to their home streams in odd and even year lines and there is a distinct genetic difference between the two lines even in the same stream. This means that while there are about three thousand pink salmon streams in Southeast Alaska, there are really over six thousand separate populations. This difference may be the result of recolonization of the streams from different glacial refugia. More than other salmon, however, pinks stray at a higher rate from their home stream, particularly if there are other spawning streams nearby. This might be a survival tactic that evolved with this fish's tendency to spawn in small streams and it ensures that fish populations headed for an uninhabitable stream are not eliminated. Stocks are also divided into early, middle and late spawning groups with timing based on stream temperatures. Those from cooler mainland waters such as Anan Creek arrive earlier than those on the outer islands.

Research on genetic identity is still in the early stages and there are no firm answers yet about the reasons for Anan's pink salmon profusion and genetic uniqueness. One clue, however, to the stream's high numbers may lie in Anan's geography. Lakes modify extreme weather and rainfall conditions. Both Anan and Boulder lakes as well as numerous smaller ponds feed Anan Creek. These contribute to the water flow in dry times. (Yes, there *are* such times in Southeast Alaska.) Lakes also moderate water temperatures, offering incubating eggs some protection against freezing. Pink salmon prefer streams with shallower riffles and more rapid flow than other salmon. Good streambeds should have velocities less than the sustained speed of the spawning salmon and relatively flat gradients paved with gravel one-half to two inches in diameter. Anan has all of these.

The life history of pink salmon revolves through a great aquatic round. As with any cycle, finding the beginning point begs the definition of a circle so we will just jump in and be carried along in the salmon's story. Adult pink salmon enter Alaska's streams between late June and early October. They generally arrive at Anan around the Fourth of July, filtering into the lagoon then heading toward the creek and lower falls. Males generally return before the females. Before they reach the prime spawning areas above the first falls the fish must dodge a gauntlet of bears and eagles as well as hurl themselves over the churning waterfall. Once in the calm riffles and pools above the cascade the female lies on her side and vigorously beats against the stream bed, removing silt down to the firm gravel beneath. In one to two feet of water she will carve out her redd which will have a raised rim on the downstream edge. When construction is complete she will lay a part of her eggs. These fall into spaces between the gravel and are immediately fertilized by the milt of one or more males. More digging by the female covers the eggs. The process is repeated until all of her fifteen to twenty-seven hundred eggs are released. She will usually guard the nest afterward but die in a few days or weeks. Redds of light-colored gravel scallop the first four or five miles of Anan's streambed where pinks have spawned.

The eggs hatch in approximately eight to twelve weeks in midwinter and are now called alevin. They remain in the gravel and feed on the attached yolk sac, growing and developing until late winter or spring when they swim up out of the gravel. The tiny fish, their egg sac now consumed, are renamed fry and measure the length of a spruce needle. It is at this point that Anan's lagoon may play a part in the creek's strong pink population. Research is finding that migration timing is linked to blooms of plankton and water temperature so that the lagoon may act as a nursery, incubating young fry in a fresh/saltwater mix until the optimal moment arrives to test wider waters. The fry undergo physical and behavioral changes - scales become larger, color turns silvery, tails lengthen and become more deeply forked in this brackish nursery. Once transformed, they change names once more. Now as smolt they migrate during darkness into saltwater, sometimes covering ten miles in a single night. In Southeast Alaska this downstream migration generally takes place during April and May. Those fish that reach migration stage are among the minority. Even in highly productive streams only ten to twenty percent survive the freshwater period. After entering salt water, the young pinks move along the beaches, carried by surface

## Anan's fish ladder

Anan's fluctuating pink salmon run and the canneries' dependence on it have prompted various measures by fisheries managers to both gauge the numbers of returning fish and improve upon them.

A weir stretching across the creek above the first falls was built prior to 1921 to permit employees of the U.S. Bureau of Fisheries to count pink salmon as they made their way up past the first falls. Two cabins above the falls housed the men who lived there during the salmon spawning months. A dam installed in 1930 below the falls was intended to raise the water level and "afford a more favorable resting place for the ascending fish," said the *Wrangell Sentinel* Another dam and a fish ladder were constructed in 1936 and an aluminum fish ladder was installed in 1967. This ladder couldn't operate efficiently during times of high flow and was removed in 1975 and replaced with a vertical slot fishway visible under the present bear observatory. This 100-foot tunnel allows migrating fish to reach spawning areas when fish cannot make it over the falls due to high water flow and velocity.

Fish ladders and other means of assisting salmon over the falls, while well intentioned, have faced opposition from some who have feared that helping weaker fish ascend will result in negative genetic changes to Anan's pink salmon population. Fisheries biologist Richard Myron wrote in 1977 on the eve of Anan's fishway construction: "The basis of the Anan productivity is the particular genetic composition of the fish combined with a stream that contains a large volume of high quality spawning gravel and relatively stable environmental conditions. The importance of the genetic constitution of the stock cannot be overstressed...Environmental conditions favorable to a particular characteristic will tend to develop and preserve that characteristic...If, for example, a partial barrier on a salmon stream excludes fish which cannot jump or do not have the power to swim against fast currents, selection pressures will tend to eliminate the offspring of those fish in favor of progeny of adults who have the desirable characteristics of vigor and jumping capabilities." (Barrick, p. 43) He concludes that, "the locally adapted races of fish, such as the Anan population, should be kept as natural and pure a stock as possible." Myron's argument briefly slowed construction of the fishway but ultimately did not halt it.

Because snowfall is less than in earlier decades, large volumes of water seldom hinder salmon from jumping the falls these days. The fishpass is still maintained but rarely opened.

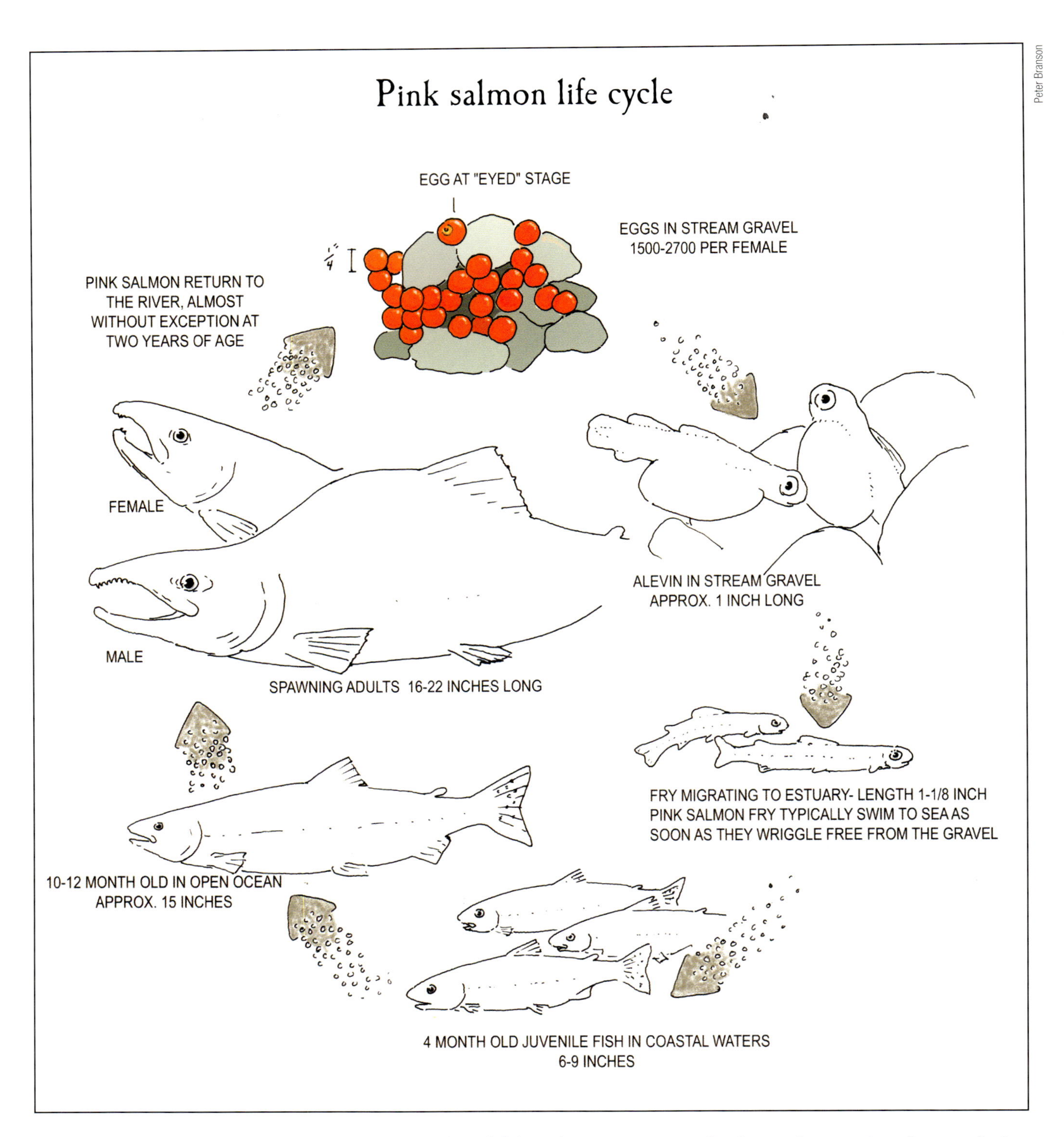


currents, in dense schools, feeding on plankton, larval fish and insects. A year after being deposited in the gravel, they have grown four to six inches long and now move toward ocean feeding grounds in the Gulf of Alaska and Aleutian Islands. Here they continue to put on weight until their second summer.

Studies have shown that pink salmon coming from specific coastal areas have characteristic distributions at sea which are identical from year to year. While at sea, they school together to thwart predators. Explanations for the salmon's ability to navigate in the open sea range from celestial to solar and sonar clues. Some scientists are now investigating whether the earth's electromagnetic current might point them toward home. Nearing their natal stream pinks, as do all salmon, tune into their ability to sense temperature changes and the chemical "smell" that distinctly identifies each watershed. Closer still, they are stimulated and drawn forward by rheotaxis – a response to the flow of opposing water current.

Pink salmon are the smallest Pacific salmon weighing in at three and one-half to four pounds with an average length of twenty to twenty-five inches. Anan pinks are generally larger than average, perhaps because the waterfalls before the spawning beds pose an extra hurdle that only the fittest can scale. A bright ocean fish is a vivid metallic blue on top and silvery on its sides and splattered with many large oval black spots on the back and the tail (caudal) fin. Its scales are very small and its flesh pink. As it approaches freshwater, the fish undergoes another physical transformation, one that earns the males their name humpback. The male's intense color changes to brown or black above and its belly becomes white. More pronounced is the shift in body shape as the male develops a large hump behind its dorsal fin. The head enlarges and the lower jaw turns up so that he cannot close his mouth. The reasons for the development of the hump may be to streamline the male's body so that it can maintain its position behind the female in a stream's fast-moving current. The humped back may also keep the salmon from spawning in water that is too shallow for egg survival. Females change also, becoming olive green with dusky bars above and a light-colored belly. Males will be the first to return to the streams and occupy the spawning grounds. At Anan Creek this will start around the beginning of

Mark Emery

*A salmon egg hatching.*

Ivan Simonek

*Pink salmon mass before tackling the falls.*

July and continue through August.

Once in the spawning stream pink salmon become prey to bears, among others. Bears have been shown to prefer unspawned females but research has shown that pink salmon respond to black bear odors with a fright response. Pink salmon are not the only fish in Anan's waters. The creek is also home to small numbers of cohos, Chinook, sockeye and chum salmon as well as steelhead, cutthroat and rainbow trout. But, it was primarily pinks as well as hunting and the level gardening ground that drew early people, including the Tlingit Indians, to the creek.

Ivan Simonek

*The pink salmon's tail spots are evident in this brownie's catch.*

✦ C H A P T E R   T H R E E

# We Have Always Lived Here

Aboriginal use of Anan's resources stretches into the distant past. Numerous weirs, fish traps and shell middens in nearby Bradfield Canal and Eastern Passage indicate that the area has been in human use for at least thirty-five hundred years. Nearly every stream in the Canal manifests subtle signs indicating the presence of early people - semicircular patterns of boulders, lines of adze-formed stakes (some still interwoven with sticks,) middens of scallop, mussel and cockleshells and the depressions of seasonal camp houses.

Evidence of human occupation of Southeast Alaska, however, dates back at least as far as ten thousand years. By that date an ice-free route between the Cordilleran and Laurentide ice sheets of the interior continent and another route down the coast allowed migration from Asia and the Alaskan interior. Sites on Baranof Island and Hecata Island, among others, support the theory of early settlement of the Western Hemisphere via coastal migration. These early people would have had to be familiar with boats and coastal resources. The characteristic Northwest Coast tradition begins to be evident around 4600 BP. Large winter villages and seasonal camps for harvesting marine resources begin to appear. Most archeological sites thus far discovered in Southeast Alaska date from 5000 BP or younger. This is because sea levels that have both risen and fallen in places by as much as 450 feet have obscured settlements that now may be covered by either water or forest. As mentioned earlier, however, evidence of human habitation in Southeast Alaska as much as ten thousand years ago is emerging.

Over one hundred fish traps have been documented in this region with an average age of nineteen hundred years. The oldest, at about five thousand years, was found on the southern shore of Mitkof Island, about forty miles north of Anan. Remains of fish traps at Anan and throughout the Bradfield Canal are more evidence that the area's abundance has attracted generations of people. Petroglyphs, those inscru-

Alaska State Library/Vincent Soboleff, photographer/PCA 1-52

*The Tlingit used many technologies to catch fish. Here "Indian Charlie" spears a salmon in Freshwater Bay, 1901.*

table messages in stone often associated with salmon streams, have been reported at Anan, though never located by archeologists. Three rock cairns on neighboring peaks may point toward Anan. A cave on the shore opposite Anan has yielded archeological, human-related dates of 3405 BP. With its wealth of food, it was inevitable that Anan and the neighboring Bradfield Canal would be a locus for early hunting and gathering people.

Anan Creek lies within the territory of the Stikine Tlingit or Stikinkwan that stretched along the mainland coast from Cape Fanshaw north of Petersburg to the midpoint of the Cleveland Peninsula and west to portions of Kupreanof and Prince of Wales Island, Mitkof, Zarembo and Etolin islands. Next to the Chilkat they were considered the most powerful and warlike of the Tlingit tribes. Several clans of the Stikine Tlingit claimed use of Anan. Much of the Bradfield Canal belonged to the powerful Nanyaayih and their Chief Shakes was reported to have a summer camp at Eagle River, a few miles east of Anan Creek. Three clans, the Teeyhittaan, Kiks.adi and Kaach.adi, owned portions of the Anan area, although there may have been others with now-forgotten claims. A bustling summer fishing village once blanketed the peninsula and shores of the Anan lagoon and bay. Gardens and smokehouses covered in red or yellow cedar bark would have been spread throughout the level areas. Elders Herbert Bradley and Matilda Marian Paul recalled that smokehouses were located on the peninsula at the mouth of the creek and on the shore opposite the peninsula. Numerous smokehouses, Paul remembered, also lined the beach terrace on the present western shoreline of Anan Bay. Thomas Ukas also remembered a big village there. Several Wrangell elders were interviewed in 1946 about their knowledge of Tlingit land use and ownership in the region. Charles Borch of Wrangell, one of those interviewed, remembered: "There were lots of community houses at Old Town and Anan Creek." [Old Town was a winter village of the Stikine Tlingit before they moved to the present site of Wrangell when the Russians constructed a fort and fur trading outpost there in 1834. Old Town was located on Zimovia Strait on the western shore of Wrangell Island.] Bradley concurred, saying that people occupied the village at Anan after the Russians arrived in the area. Willis Hoagland, also interviewed in 1946, recalled that people hunted bear up Anan Creek.

Summer houses were smaller and more roughly built than the massive winter houses. They were covered with bark or planks and often without floors. Sometimes the same building would house a family and their smokehouse together. Archeologists have tentatively identified house depressions, garden areas and unusual squared "storage pits" on the peninsula near the creek mouth, although no formal excavation has taken place. People also used to occupy a village at Marten Creek on the north shore of the Bradfield Canal opposite Anan Creek, added Paul. "The Marten Creek people fished and planted gardens at Anan Bay." (Anan Creek History)

"The Natives planted carrots, rutabagas, turnips, and potatoes at the camp," Paul remembered. By the time of their interviews people no longer lived at Anan Creek. They did, however, still come to harvest

Clarence Leroy Andrews /Alaska State Library/Historical Collection/PCA 45-124

*This view of a Tlingit summer village in Funter Bay, taken in 1915, gives some clue to the way Anan's summer village might have looked.*

# The Tlingit Cycle of Food

*The Tlingit annual cycle of food collecting began in March as days lengthened and stored food diminished.*

**March** - *Moon when sea flowers and other things grow in the sea.* The month for halibut and cod fishing, trout fishing in streams. Clams and mussels are dried, smoked and packed in boxes. Trap for marten and mink. Brown bear in dens marked in autumn or winter are hunted with dogs and spears or pits and deadfalls.

**April** - *Moon when real flowers grow.* Catch and dry halibut, fish for king salmon, hunt grouse, gather seaweed.

**May** - *Moon when people know everything will grow.* Continue to fish for halibut and cod. Gather spring greens, dig and store roots and bark of hemlock. Fish for herring and eulachon for their oil. Gather herring eggs.

**June** - *Moon of king salmon.* Fishing and hunting is done for direct consumption. Collect sea bird eggs.

**July** - *Moon when everything is born.* Salmon begin to crowd the mouths of rivers. Gather fresh salmon eggs.

**August** - *Moon when things get fat.* Storage activities begin. Women go out for berries, salmon eggs are mixed with berries and stored in eulachon oil. Deer, mountain goat and black bear hunting begin.

**September** - *Moon when fish and berries fail.* Salmon are caught, split and dried or smoked, hunting for goats and bears continues. Go to potato grounds and dig potatoes

John N. Cobb courtesy University of Washington Libraries/ Special Collections/Cobb 3369.

*Seaweed drying in cakes at Shakan, Kosciusko Island, 1911. The Tlingit carefully prepared the seaweed before storage.*

**October** - *Big moon.* Hunting continues. Assemble in winter villages.

**November** - *Moon of snow shoveling*, ceremonials, feasting and storytelling

**December** - *Moon when animals in the womb begin to have hair.* Ceremonials continue, preparations for tool making, canoe and housing building. Carve, weave baskets and make repairs.

**January** - *Moon when sun and geese return.* Carve, weave baskets and make repairs.

**February** - *Black bear month.* Young are born. Prepare fishing gear.

*Compiled from Oberg and W. Olson.*

Bonnie Demerjian

*This small peninsula in the Anan lagoon is the site of the former Tlingit village and the Sailor Fishing and Mining Company.*

berries and other foods. Dick Stokes, Wrangell Tlingit elder, used to visit there as a boy with his grandparents. Even when ice was still on the creek, they fished for steelhead trout. In June the family camped on the sandy beach at the mouth of the bay while gathering herring that spawned on kelp and hemlock branches. In summer they picked thimbleberries and fished for salmon in the creek and crabs and halibut in salt water. By the time he was seven or so, Stokes began to fish commercially with his uncles, still following the traditional Tlingit way of educating boys under their uncles' tutorship.

Bonnie Demerjian

*Anan's lagoon at high tide. At low tide much of the water empties out.*

Ownership, more accurately trusteeship, of a summer site was often based on the tradition that a great-uncle or some other clan ancestor had "discovered the place". Possession was founded on a story that explained the name of the place. The name Anan, sometimes spelled An-An or An'aan (pronounced Ahn ahn in Tlingit), is said to mean Sit Down Town and may come from the fact that in the morning people would go out of doors and sit down to talk things over. Although few visitors today take the time to sit down (and the number of bears in salmon season might make even a short "sit down" less than completely relaxing) the name conjures up an image of plenty in a culture confident of itself. Because of the creek's documented importance in the story of the Stikine Tlingit, a portion of the small peninsula in the lagoon has been deeded to Sealaska, the Southeast Regional Native Corporation formed as a result of the 1971 Alaska Native Claims Settlement Act (ANCSA).

No food was more important to the Tlingit than salmon and because of that, the salmon run dictated their economy and calendar. In late spring or early summer clans moved to their fish camps and remained until autumn. There they ate fresh salmon boiled, baked in an earth oven or roasted on a spit. The women gutted and cut the fish, each woman having her own distinctive pattern. The fish to be stored for winter was soaked in brine briefly then hung on poles, smoked for three or four days and finally baled in bundles. The heads were tried out for grease, using a small canoe filled with water and heated stones. The fish were boiled until the grease rose to the surface where it was skimmed. Fish entrails were returned to the stream or burned to ensure renewal. These skilled and laborious tasks all fell to women but, in turn, control over this important food supply helped lend women their generally high status in the household. The skill exercised in salmon preparation and storage extended to all other foods. Plentiful food was instrumental in Tlingit survival and in the development of their high level of social and artistic culture.

The Tlingit held salmon in deep regard and an elder from Angoon illustrates this. She says, "One person was delegated to be responsible for the fish. Every day, he watched the ocean beach for fish jumps and kept track of all movements of the fish. No one was allowed to kill fish before they came upstream to spawn, they believed if the fish was bothered and disturbed during their migration upstream to spawn, they would turn back and go up another river. Since fish was our main food, we were very careful; the fish were treated well. If a man broke one of our laws, his fishing equipment was taken from him; sometimes his spear was broken." (Newton and Moss, p. 4) This conservation practice aided escapement, a practice that the first white fisheries ignored until the near devastation of the fishery.

To the men was assigned the business of catching the salmon. There was no standard model of fishing technology. Instead, the Tlingit considered topography, water current and other natural features, using all to advantage. Various types of traps and weirs were used as well as nets and spears. Traps and weirs made use of the salmon's instinctive drive to head upstream. One method was semicircular traps of stones that collected fish into boulder-lined basins. Fishers used rising tides to usher the salmon into the traps. High water brought fish congregating at a stream mouth over the upper edge of a trap on the flood tide. Then as the tide receded salmon became stranded in shallow basins that kept the fish alive for a time until they were harvested.

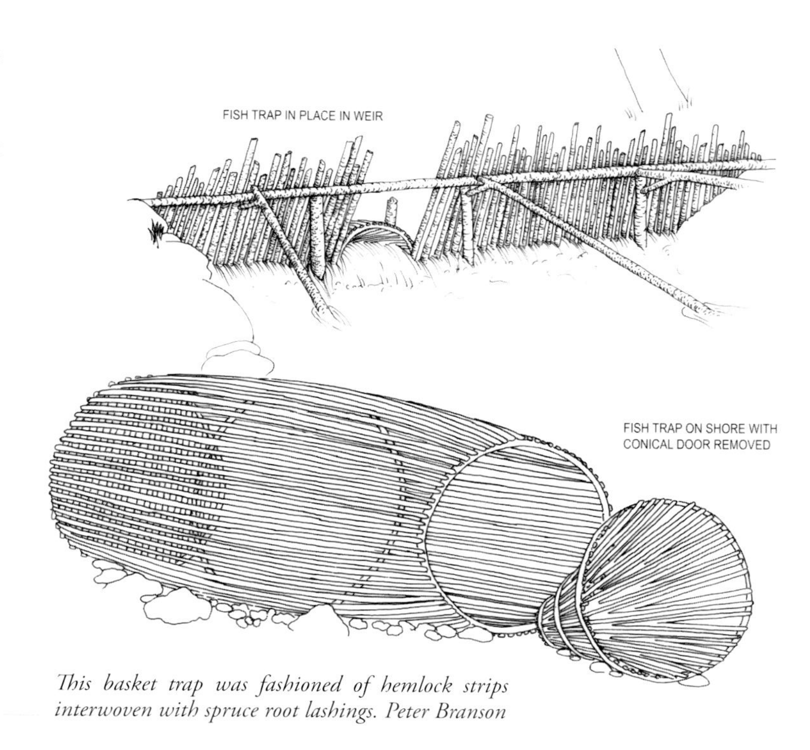


*This basket trap was fashioned of hemlock strips interwoven with spruce root lashings. Peter Branson*

Stone weirs or walls were also built in the intertidal zone. Gaps in these obstructions were blocked with baskets woven of laths and lashed with spruce roots. Using the salmon's drive to swim upstream, the traps caught the fish as they swam through the gaps into the basket from which they could not exit. Another weir technique consisted of lines of wooden stakes, often of Western hemlock, embedded in the gravel and sometimes laced with lattice fences. These also herded the fish toward a trap. Occasionally salmon were caught with gaff hooks having a detachable hook like a harpoon. These were used from the bank of a stream. Nets were also used, first of twisted cedar bark, later of commercial twine once it was introduced by Europeans.

The Tlingit were selective, asserts Alaskan anthropologist Steve Langdon, in providing for sufficient escapement to maintain healthy runs. They did this by positioning their traps and weirs at the half-tide mark in the intertidal zone, ensuring that at high tide the structures would be covered with water. Thus, at high tide the fish would be free to swim upstream unimpeded. Only those that did not make it all the

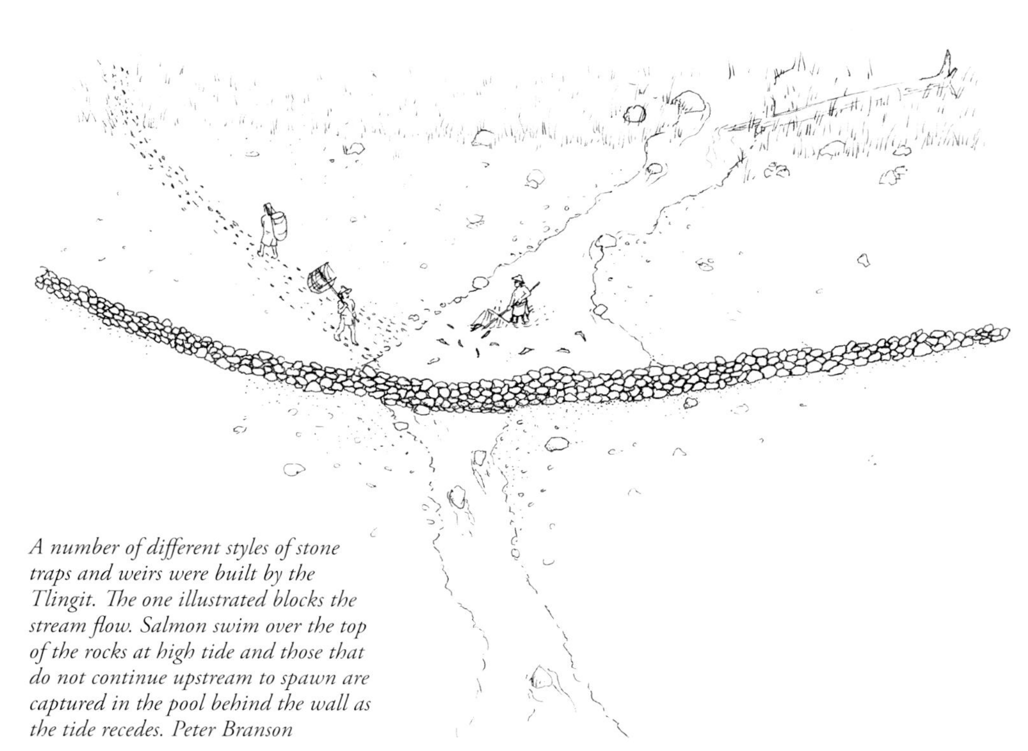

*A number of different styles of stone traps and weirs were built by the Tlingit. The one illustrated blocks the stream flow. Salmon swim over the top of the rocks at high tide and those that do not continue upstream to spawn are captured in the pool behind the wall as the tide recedes. Peter Branson*

# Tlingit Salmon Myths

*There are several Tlingit myths that tell about how this staple food came to be.*

### Raven Creates Fish

One day, Raven called the salmon together to choose their rivers. King Salmon said "I will travel up the long, large rivers to the clear waters, where I will spawn." Dog Salmon was next. "I, too, want the larger rivers," he said, "but if they are filled, I will use the smaller streams." Next came Coho. "I prefer the short, fast, clear waters for my spawning." At the end was poor little Humpback. He looked up and softly said, "I'll take whatever is left." And so it is that even today, each type of salmon can be found in the streams they picked. But you will notice that there are more Humpback salmon than all the others.

— adapted from a legend told by Walter Williams, Tlingit storyteller, in *The Tlingit* by Wallace Olson

Ivan Simonek

*A little interspecies conversation, perhaps?*

### Fog Woman

This story comes from the days there were no salmon and Raven had to eat other fish. One day, Raven, who was camped at Anan Creek, went out in mid-channel to fish. A heavy fog came in and he could not see. Then a woman appeared in his canoe and asked for Raven's hat. All the fog poured into it and the sky became clear. Raven took this woman for his wife.

One day when Raven was away Fog Woman sent one of the slaves to get water in his hat and when he returned there was a salmon swimming in it. This was the first salmon and they cooked it and ate it before Raven returned. When he came home he smelled the fish on the slave's breath and found out that Fog Woman had created the salmon. From then on there were many salmon that they caught in the stream and smoked. Soon after Raven began to forget that it was Fog Woman that had brought him his good fortune. They quarreled more and more. Then finally in a temper Raven threw a piece of dried salmon at her. She ran away from him, but, when he ran after her and seized her, his hands passed right through her body. Then she went into the water and disappeared forever, while all of the salmon she had dried followed her. After that he went to his father-in-law to beg him to have his wife come back, but his father-in-law said, "You promised me that you would have respect for her and take care of her. You did not do it therefore you can not have her back."

Some say that Fog Woman's daughters, the Creek Women, live at the head of every stream and the salmon each year fight their way back to see her. All die attempting this except the steelhead trout who comes back every year.

— This story is told on the Chief Johnson totem standing in Ketchikan.
(from Keithahn, *Monuments in Cedar*)

way upstream would be caught on the ebb tide. Some Tlingit groups were also careful not to block the channel at the high water range. Langdon estimates that Tlingit traps operated less than twenty-five percent of the time that salmon were in the area, ensuring that runs would continue to be replenished each year.

The Tlingit believed that humans were intimately linked with animals and shared qualities with them. Salmon, for example, like humans, were a powerful tribe divided into five clans that lived somewhere in the ocean. According to a legend reported by anthropologist George Emmons, their country, far away to the westward, was surrounded by an ever opening and closing ring, through which they had to jump quickly to preserve their clean silver sides. Those salmon that were caught were cut or marked, which accounts for the stripes on their sides. Salmon were believed to travel in invisible canoes. The chiefs of the different families stood in the stern to direct their movements landward. When spring came, a great meeting was called and all wanted to start at once for their streams. The dog salmon, through jealousy, broke up the canoes of the cohoes, which accounts for the late appearance of the latter and also for the sequence of salmon species arrivals each year. (Emmons, p. 104))

As with the invisible canoes of the salmon people, those of the Tlingit became harder to see. With the advent of commercial salteries and fish canneries in the late nineteenth century, the unsettling influences of gold and fur and the ravages of disease, the traditional Tlingit way of life, woven from the threads of hunting and gathering, faded from its former vitality. In 1899 the leaders of the Wrangell clans appealed to John Brady, the territorial governor of Alaska in a moving letter. In it, George Shakes, Tomyatt and other leaders pointed out their dependency on salmon streams and the depredation that canneries were already inflicting on the salmon and the people. "We only ask for justice according to the laws of the United States. We are very thankful that the Government of the United States has such a law as to protect us with our lands and salmon streams, especially salmon creeks and hunting grounds. For all our living and our children's living comes from fishing and hunting. And so if the white man takes the salmon creeks away from us, where do you suppose, Governor, we Indians would make our living – we old Indians...that cannot talk a word of English or not able to work for white man for wages due? They simply have to starve or else look to the Government for support." The letter goes on to describe a friend, Aarow Kohwow who owned a place called Au-Aw, meaning the Town of all Towns and one of the oldest towns in the area. It states that J. Babler, superintendent of the cannery took fish from the creek there, broke down a garden fence and tried to keep all others away. "There are three good creeks where we all use to fish: now one white man has it all in his possession and puts a fence across each creek every year, and we Indians left out...Let us have the creeks where we get our living and hunting grounds: that is all we ask of the Government." (Ibid. p. 6) Jefferson Moser, who made a survey of Alaskan salmon resources for the federal government in the earliest years of the twentieth century, argued in the same report for fairness in dealing with the Native people but saw a pattern being repeated in Alaska that had already brought disaster to aboriginal people elsewhere: "From the Indians' standpoint, their complaints are undoubtedly well founded, but history will no doubt repeat itself here, as in other portions of our country, where the aborigines have come in contact with the civilizing influence of the white man, where rum, disease, and mercenary dealings have slowly but surely exterminated them. My own sympathy is with the Indian, and I would gladly recommend, if the way were clear, the establishment of ownership in streams: but it is impracticable, and I can only ask for him a consideration of his claim and, whatever law is framed, that a liberal balance be thrown in his favor." Moser's pleas and those of Wrangell's native leaders were noted, filed and ignored. (Ibid. p. 4)

The unimaginable abundance of salmon was to have a deep impact on the Tlingit people who were as vulnerable as the fish and just as intimately linked to the salmon streams. The anadromous nature of the fish makes them more susceptible to overfishing than most marine fish, since they can be harvested near their natal creeks when they are densely congregated. As salmon streams in California and the Pacific Northwest were decimated, eyes turned north to Alaska's salmon. The Tlingit themselves became nearly invisible.

# CHAPTER FOUR

# We Feed the World: Commercial Fishing at Anan

Throughout the history of commercial salmon fishing in Alaska runs a consistent note of anxiety about the depletion of the resource. The salmon of California and the Pacific Northwest had already been well exploited as the result of reckless and improvident fishing when the first Alaskan salmon cannery was built in 1878 in Klawock. This was preceded by Russian harvest and salting of sockeye salmon for subsistence use and export. By 1889 there were thirty-seven canneries in operation in Alaska but the expansion was too rapid for the market and by 1892 the number had fallen to fifteen. This small number was in part the result of the consolidation of a number of independent canneries. Many companies were bought out not for the canneries themselves, but for the fish trap sites that they controlled. The resource exploitation of the resource was unrestricted and the race was on.

The Aberdeen Packing Company was the first cannery in the Wrangell area. The company built a plant in 1887 on Point Gerard situated on the mainland about eight miles above the mouth of the Stikine River. After two years, the operation, under the name of the Glacier Packing Company, moved across the channel to Point Highfield on the northern tip of Wrangell Island and in 1893 it joined the Alaska Packers Association. Besides the Stikine River and other locations, the cannery processed fish from its fish trap Number 20 near the mouth of Anan Creek. The cannery supplied gill nets, lodging and board to fishermen who were paid fifty cents per one hundred humpbacks. In 1897 the cannery took 375,000 Anan humpbacks. Other canneries, including the Point Warde cannery nearest the creek, another

J.E. Worden, Wrangell Museum.

*The Alaska Packers' Association (APA) cannery at Pt. Highfield in Wrangell in September, 1914. Note the ship, one of APA's Star fleet. This photo would have been taken just before the ship headed south with the summer's salmon pack*

in Santa Ana Inlet and in Union Bay also reaped Anan pinks.

After 1890 the use of salmon traps spread rapidly to increase catches and insure greater profits. The first commercial traps were constructed in the spawning streams themselves and were deadly efficient. These fences permanently barricaded entire streams and consisted of anything from simple brush barriers to heavy tree trunks and piling fences braced by smaller logs and hung with webbing. These were placed at or above high water and so trapped all spawning fish. Once the salmon were schooled, they could be easily dipped out. Portable traps built of wire net called chickenwire, popular in Southeast Alaska, could be rolled up and carried from stream to stream. Even at this early date, observers such as Jefferson Moser voiced concern about declining salmon stocks. He stated to Alaska's governor in 1899, "the causes of the depletion are the barricading of streams and overfishing." (Governor's Report 1899 p. 4) Congress banned dams and other obstructions in 1889 and in 1896 forbade fishing above tidewater in streams less than 500 feet wide but there was no funding for management or enforcement until 1898. Then, two agents were hired but given no transportation money. One Alaskan complained in the *Wrangell Sentinel* that the government's supervision of traps was "like trying to hunt big game with a brass band". (January 16, 1919) The agents' few scattered gas boats were spread over thousands of miles of water. This pattern of passing legislation without supplying the means to implement it continued for decades.

W. H. Case, Wrangell Museum

*"The APA Co. trap crew, An-An Creek, Alaska" These men maintained and worked APA's trap No. 20 that caught salmon returning to Anan Creek.*

Pile-driven traps evolved from similar but smaller Native traps. These took advantage of the salmon's inclination to migrate along shorelines to creek mouths and then congregate there before heading upstream. Hand-driven stakes hung with net directed salmon into the trap. Pile traps were costly since the pilings had to be removed and re-driven each summer. They were also devastatingly efficient. In 1903 a pile trap near the mouth of Anan Creek, probably belonging to the Sailor Fishing and Mining Company, took an estimated two million pink salmon.

As described by an early admirer of the technology: "It is most simple in its construction, and consists of a long arm of piling and netting reaching out at an angle to the sea. The fish are stopped by the net, which is fastened to the piles and extends to the very bottom of the water. Continuing their way up against the trend of the water they pass through a narrow funnel that opens into the trap proper. The trap is completely covered on the bottom with a great net and the fish, crowding through the opening, find themselves in a trap from which there is no escape...This immense net is lifted from the inside of the trap at stated periods and the catch is dumped unceremoniously into waiting scows." (Colt)

Alaska State Library/Case & Draper, photographers/PCA 39-717

*APA trap No. 20 at Anan was first placed close to the creek, then was required by law to move farther and farther away to allow more salmon to enter the stream.*

By 1906 no such fixed gear was allowed in rivers or narrow bays although it was not banned in mouths of open bays such as Anan until years later. The *Wrangell Sentinel* spotlighted concern about these traps when it

# The Point Warde Cannery

"A new cannery at Point Ward[e]...is assured." The *Wrangell Sentinel* in September 1911 quoted W.D. Grant, a Wrangell pioneer, trumpeting the immanent construction of yet another cannery in the area. The Point Ward Packing Company, in a diminutive, sheltered and unnamed cove just around the corner west of Anan Bay would target Anan pink salmon and "red fish" or sockeyes nearby. A crew of fifty rapidly built a "strictly modern, two-line cannery of thirty thousand [cans] per day," and the operation began production the following spring. Promoters hailed from Chicago, Oregon and Juneau with Grant, a local man, chosen as superintendent. This new packing company was typical of the small plants that were springing up around Southeast Alaska.

Prior to the establishment of the cannery the cove was a Tlingit camp. In 1946 a Tlingit elder remembered that it belonged to the Xook'eidi clan and was used for trapping marten. An old Forest Service map shows that a fort, presumably Tlingit, was once located on the site but perhaps what the mapmaker saw were remnants of previous trapping camp shelters. Grant and his associates recorded a trade and manufacturing site in 1911, permitting them access to this land.

Wrangell Museum

*The Pointe Warde cannery around 1912.*

Captain George Vancouver had given Point Warde its name while exploring the area in August of 1793. The Point Warde cove is scenic, graced with a modest but reliable stream of water falling across rocks into the bay. Because of the scarcity of flat land, rock on the hillside had to be blasted to make room for the buildings that also extended on pilings out over the water. A large canning room, fish house, warehouse, bunkhouses and powerhouse would all have been necessary to house this self-reliant community. Six scows and twelve seine and gillnet skiffs were also built to catch and haul fish to the cannery.

Point Warde cannery packed both trap and seine-caught fish from 1912 to 1929. Under the Sprite label the cannery put up 34,000 cases its first year. The cannery took fish from its trap until new regulations made it illegal to fish so close to Anan Creek. Other packers also were forced to move their traps further from the creek and closer to nearby Point Warde.

In its second year the cannery made headlines in the *Wrangell Sentinel* when it was robbed during the winter, a crime abetted by its assistant watchman. Two men broke into the warehouse and stole net and other gear. They were "caught with the goods" and confessed, losing no time in accusing the assistant watchman in response.

Wrangell native Dick Stokes recalls living at the cannery as a small boy. His parents worked there - his mother in the cannery, his father operating a seine boat. The family lived in one of the small houses built for workers on a terrace above the bay. Although he was only about four years old, Stokes has a vivid memory of that time because of an accident that might have proved disfiguring. He reached up to the stove and pulled a pan of hot boiling potatoes onto himself. Stokes was hurried back to Wrangell where his grandmother doctored him with a home remedy that included cottonwood bark.

In 1920 Point Warde and a large number of other Southeast canneries merged. Several other companies leased the cannery during that decade and operations finally ceased in 1929. The next year much of its larger equipment, including the large cylindrical retorts in which the cans were cooked, was dismantled, probably to be reused elsewhere. Its lumber was left for salvagers. Today curio seekers find ruined pilings, wooden stave pipe for carrying water and the odds and ends of cannery life scattered through the mossy woods. The inlet's local name, Cannery Cove, is a faint reminder of its once lively past.

## The Sailor Fishing and Mining Company

The first commercial use of Anan's salmon wealth is recorded in articles of incorporation in the state of California in 1897 for the Sailor Fishing and Mining Company. The articles stated grandiosely and exhaustively that the corporation intended to "catch, can, salt, dry, cure, preserve, buy, sell, import and export fish and the products of the sea", to operate vessels "in the United States or anywhere", acquire land in Alaska or anywhere, and to "prospect, explore, develop, work and mine any and all ledges". Each of the five directors subscribed fifty dollars toward this grand ambition. They filed a homestead claim at Anan in 1899. A small .69-acre plot on the lagoon's peninsula was surveyed that year and in 1900 the Department of the Interior issued a plat for a "homestead claim on the right bank [According to standard practice, the "sides" of streams are designated according to descending position.] of An-An Creek on Eastern Passage" to the corporation.

In 1862 and 1864 Congress had passed an act "to secure homesteads to actual settlers on the public domain" in order to fill up the nation's new acquisitions with American farmers. These homesteads were eighty acres in size. The organizers of the Sailor Fishing and Mining Company, based in San Francisco, applied for their patent under the later Soldiers' Additional Homestead act of 1872 that granted those who had served in the Civil War an additional eighty acres. The Sailor Fishing and Mining Company was, in fact, a shell company for the large conglomerate, the Alaska Packers Association, which was attempting to amass trading and manufacturing sites in Alaska at this time. The Anan homestead plat shows a house and fencing around the perimeter of the peninsula and a pile trap was planted at the mouth of the bay. Despite its name, however, there is no evidence that fish processing or mining ever took place on Anan Creek.

In 1901 the company built a trail along the lagoon to support its salmon harvesting activities, possibly improving a bear trail or one used by the Tlingit. The federal government took over maintenance of the trail in the 1930s long after the packing company had disappeared. The Alaska Packer's Association (APA), bought the plot in 1905 for ten dollars in gold coins and in 1914 the Sailor Fishing and Mining Company ceased to exist.

The APA donated their holdings at Anan to the Forest Service in 1942 as a public recreation site. That site and another on the creek shore opposite it have now been claimed as historic sites under the provisions of the Alaska Native Claims Settlement Act (ANCSA) and have been conveyed to Sealaska Native Corporation. Federal law protects this land from mining and other commercial activities.

reported on a petition circulating around Alaska to ban fixed traps and stated: "The number of these fish traps is increasing so rapidly that they will eventually throw most of the fisheries out of employment." (June 18, 1908) William L. Paul Sr., Native lawyer and activist, reported to the Bureau of Fisheries in 1905 that no salmon were getting into Anan because of a trap at the creek's mouth. He was still protesting traps in 1941 and urging sufficient escapement when he set his seine directly in front of Anan Creek after first notifying the Fish and Wildlife Service that he would be doing so and inviting them to arrest him. Dick Stokes, deckhanding on a tender nearby, boarded Paul's boat and helped with the set. No one showed up to interfere, recalls Stokes.

In 1912 the *Wrangell Sentinel* (October 24, 1912) reported that commercial fishing would soon be prohibited in Anan's creek, lagoon, lakes and tributary waters and within five hundred feet of its mouth, but the first territorial legislature killed the measure. In any event, federal law prohibited any legislative body from passing laws relating to Alaska's fisheries That same year the U.S. Congress itself had passed the Organic Act of the Territory of Alaska giving the federal government the power to control the territory's fish, an exception to the precedent that new territories be given some measure of authority in managing their fisheries. Few federal resources for managing Alaska's were allocated, however.

The fish processing industry, ever inventive, adapted to increasing public disapproval and the high cost of the fixed traps. In 1907 the less expensive floating trap was devised. This type was constructed of floating logs anchored in strategic positions in deep water to the sea floor. Wire hung from this frame. Like the pile traps, a fencelike lead extending seaward directed salmon into a "pot" where fish were held until the cannery needed them. They were then brailed or scooped into waiting scows for transport to the cannery. In winter, the traps could be towed to sheltered bays for storage. Canneries continued to use a few pile traps in open waters of Southeast where the floating traps might break apart in foul weather. They were also employed extensively in western Alaska.

The number of traps began to greatly expand after a fishermen's strike in 1912 demonstrated to canneries the advantage of owning their own traps rather than relying on gill and seine-caught fish. Demand for canned salmon swelled during World War I and drove the number of traps to a high of 653 before a glut on the market forced a decrease in production. The number of traps, both pile and floaters, rose again, however, and by 1927 was at a high of 799 in the territory. Besides offering companies more control over their supply of fish, traps were more economical because fewer employees were required. At the height of the trap era around 1600 men were employed as watchmen, tenders and guards while in 1970 as many as 15,000 gillnetters and seine fishermen made a living harvesting salmon. Fish traps were indeed more efficient catchers of fish, and that was the problem.



*There was no standard salmon trap but in general the lead diverted the fish moving along shore into the heart or hearts and then into the pot. The fish kept swimming forward through the tunnel to the spiller. This had a moveable floor that could be raised by manpower or steam to spill the fish into the receiving vessel.*

The first seine boats operated entirely with manpower. The oarsmen, called "boat pullers", rowed to where the pinks were schooled up, set their net by hand by attaching one end of the net to the shore, then circled the school and pulled in the net using only strong backs and arms. Lots of "pushing and cussing" accompanied the latter, recalls Harry Sundberg, former Wrangell cannery operator. Later, larger seiners, still relying on manpower, used purse seine nets.

Early purse nets were hauled by hand onto the back deck of the larger boat. As many as seven crewmen, described as "successors to the galley slave", were needed to bring aboard these nets. Invention of the power block in the early 1950s changed the fishery. This hydraulically–driven wheel was attached to a boom. The seine net with its cork and lead lines was pulled through the block as men waited on deck to stack it. The power block, along with other power-driven improvements, made purse seiners almost as efficient as fish traps in years when salmon were not plentiful. For example, in 1959, William Willard harvested a legendary 55,555 pinks in *ARB No. 6* on opening day at Anan Creek.

*This photo was taken in 1914 by one of the two cannery workers, Frank K. They were stationed at Kassan on Prince of Wales Island. On the reverse he wrote to his friend Charlie who was working at the Point Warde cannery urging him to come to Ketchikan for the winter where there would be "lots of girls, spring chickens."*

Today in seining, one end of the purse seine net is attached to a small power boat called the seine skiff, the other end to the seine boat. When

## Opening Day

It would be no exaggeration to say that Southeast Alaska was built on fish. Prior to the incursion of large-scale logging operations in the region in the 1960s, fishing was *the* means to make a living. The energy that fueled Southeast communities was fish. Since Anan Creek had legendary pink salmon runs, it attracted fishers not just from Wrangell but from around the region. Opening day, the first day that seiners could throw in their nets, was an occasion for fishing frenzy. That day might occur anywhere from early July to early August but when it arrived, the boats were ready and waiting. Observers from U.S. Fish and Wildlife monitored the creek and adjacent shoreline, checking to ensure proper escapement into the creek before allowing operations to begin. A signal flare or shot triggered an explosion of sound. Fisherman Chuck Petticrew recalls the melee, beginning with a ringing clang as spring-loaded releases set the power skiffs free. "There was no time to untie from a cleat. You had to get free instantly," he explains. At the same time, diesel engines on both the seine boats and their smaller skiffs roared to life as the skiffs took off with one end of the seine nets. "It was an explosion. Ricochets of sound. You could hear it echo off the mountains," he says, as every boat clawed for a good setting position. In short order, boat and nets were jumbled, one "corking" [setting the net to intercept the fish ahead of another boat], and the frenzy was on. Opening day might, as in 1951, draw as many as 160 seiners to the pink horde.

*This advertisement appeared in the* Wrangell Sentinel *on August 12, 1949.*

Fishing was not just for the fishermen (and they mainly *were* men. Some skippers, following ancient wisdom, considered women aboard bad luck.) Sylvia Bahovic, though, was one woman who did work onboard and she remembers: "People from Wrangell would pack a picnic and go down. The town was almost deserted" as everyone who could buy or beg a ride went to see the spectacle. Others flew down in planes. One pilot remarked in 1951 that a control tower was almost necessary to handle the air traffic. Cannery operator A.R. Brueger and his wife hosted breakfast on their boat each year for visiting fishery and government officials who enjoyed a ringside seat for the event.

Opening day provided the excuse for family vacations, too. It was, after all, the best show around. Tents filled with wives and children lined the sandy beach of Anan Bay. After the main event, there was always berry picking, swimming and snagging fish.

Fisheries are more diversified today. Canned pink salmon's appeal has declined as other species and methods of preservation take over. The heyday of canned pink salmon and the circus atmosphere of opening day belong to Anan's and Alaska's past.

*An early purse seiner from 1917 when the net was still hauled into the boat by hand. Courtesy of Alaska State Library/ Alaska Purchase Centennial Commission Collection/PCA 20-157*

salmon are spotted the skiff operator releases the skiff from the seiner and tows the net away from the larger boat. The seiner then pulls its end into a hook shape and holds it, positioned so that the salmon will swim into the net. On signal from the skipper, skiff and seiner close up the net by driving toward each other. Once they meet the net is closed or pursed at the bottom so the fish cannot escape. Then the net is hauled through the power block on board. Seiners harvest most of the pink salmon since the fishes' tendency to form large schools lends itself to being rounded up in a net. Gillnetters and trollers catch smaller numbers of pinks because, though there is some overlap, different fishing methods generally target different species. The seine boat has traditionally dominated Anan Creek, where pinks outnumber all other species.

*Native fishermen shore seining in the early days of the 20th century. The canoe is outfitted with oarlocks and oars as well as with traditional paddles. Alaska State Museum/Historical Collections/ASL-Sitka-Indians-31.*

From the first, the majority of Alaskans opposed fish traps. Early canneries hired some Natives but many workers were imported from Seattle and San Francisco. Initially, canneries also purchased fish from Native fishermen. The anthropologist George Emmons stated in 1905 that about twenty-three hundred out of a total Tlingit and Haida population of six thousand were employed by the salmon industry. Around 1900, Natives began to abandon their small settlements and move to towns and scattered bays where they worked in and fished for the canneries.

The Natives had little protection against whites encroaching on their traditional fish creeks and favorite shorelines. In 1907 presidential proclamation removed the possibility of expanding the lands they had managed to keep when the Tongass National Forest was created. This put under federal control all lands not in private hands. The increase in the number of fish traps proved to be especially devastating for Native people who not only had lost control of their salmon streams, but also had become dependent on a cash economy. Native fishermen, however, vigorously continued their fishing tradition by seining pinks. Others adjusted to the tyranny of fish traps by turning to the troll fishery for kings and cohos. At the same time, as traps spread, Natives began to join the chorus of protest to the decimation of their salmon resource. Discontent on all sides centered on ownership of the fish traps. Absentee owners, the "Fish Trust", who controlled the canneries, fish tenders and traps came in for increasing wrath from most territorial residents who tried repeatedly to get traps banned. Despite continued but ineffectual federal legislation curtailing traps and fishing seasons, pink salmon harvest levels continued to fluctuate alarmingly. Fish traps became a symbol of

## Following Fish

With the rapid growth of commercial salmon fishing in Alaska at the beginning of the twentieth century came increasing uncertainty about the size of each year's run. Would it be large enough to support the canneries? Were the continuously working fish traps depleting the run? Some fluctuation was assumed to be from natural causes but the ballooning harvest began to raise the fears of Alaskans. As difficult as it is to believe today, in the 1920s it was not known whether pink salmon returned to the streams from which they had come.

In 1929 Dr. Frederick Davidson of the U.S. Bureau of Fisheries began studying pink salmon in Anan Creek and Olive Creek (also called Snake Creek) on Etolin Island, to answer this question. The Bureau expected to use the information gathered to help regulate escapement and thus protect the runs from overfishing.

Davidson clipped the adipose fins of young salmon, then watched for their return in the following years. He trapped the mature fish in weirs, examined and returned them to the creeks to spawn. He also studied the scales of pink salmon in the two creeks to determine their age. As with trees, scales are laid down in concentric rings that mark the rate of growth. In spring and summer the fish grows more rapidly than winter. This phenomenon was first described in 1912 and Davidson confirmed it. In a final report to the *Wrangell Sentinel* in 1933 he warned that the intensity of the commercial fishery "has contributed to the variability and size of the pink salmon run." What had been obvious to common sense was now proven by science, though it would take decades longer to translate his warning into meaningful change. Dr. Davidson's research continues to be referenced in salmon studies today. (Bonar et al.)

## Lawrence and Sylvia Bahovic—a seiner's life

You couldn't call it a crime. Alaskans never did, anyway. In the fish trap wars, Alaskan fishermen thought of themselves as Patriots. The other side were the Redcoats and, while this cat and mouse game had something of the romantic about it, both sides were playing for keeps.

Lawrence Bahovic began his fishing career in the heyday of expanding canneries and multiplying fish traps in 1934. His first sets were from a skiff powered only by oars. Tying one end of his net to a tree he circled the salmon then, with three other men, hauled the heavy cotton net over a roller at the boat's stern onto the deck. First, the power roller reduced this backbreaking work, then came the power block. "I had one of the first power blocks in town," he recalls. His first seine boat was a communally owned affair, called the *Lalowa*. That tongue-turning name was a concoction of his name and that of his two partners, Lloyd Benjamin and Warren Gartley. They built the boat, fished it together for a time and occasionally took advantage of Benjamin's friendship with creek watchman .30-30 Jack to harvest a few of Anan's fish. "He asked .30-30 if we could get in [closer than the legal distance to the creek mouth] to make a set [circle the seine around a school of pinks]," Bahovic relates. "He didn't answer. He just said, 'One day I'll be gone', so we knew he wouldn't be around to see us."

Lawrence remained a fisherman while the two partners moved on to other work. He moved up to skipper a cannery tender, the *XL* and usually played by the rules, hauling fish from traps to the A.R. Brueger cannery in Wrangell. But who could resist a bay full of unguarded pinks, particularly when the fine was only twenty-five dollars? Alaskan attitude toward fish piracy was benign. Contrast the *Wrangell Sentinel*'s reporting of the accident that ended the *XL*'s days with Bahovic's retelling of the same story. He chuckles as he remembers scouting out a particularly fine school of pinks in Seclusion Harbor on Kuiu Island at high tide. When the tide fell, he sneaked back to capture the fish and had about 15,000 on deck when a patrol plane unexpectedly appeared overhead. In his haste to exit the bay, he smashed into a rock. With a hold full of fish, the boat failed to right itself and rolled, demolishing pilothouse, mast, boom and bulwarks. The crew made it to shore and set up "a real jungle camp" while awaiting help. The *Sentinel* repeats these same colorful details while omitting the fact that the catch was illegal. Ever the gentleman, Lawrence publicly thanked those who rushed to help with pumps, empty oil drums and salvage aid. He later rebuilt the *XL* and renamed her the *Aurore Marie*.

Lawrence's wife Sylvia accompanied him on many trips. She remembers the day she "burned buns" twice. First were her own when she slid across the deck and sat down in a mass of stinging jellyfish brought aboard with the net, then again when she charred biscuits she had baked for the crew's dinner.

There were plenty of good times in between the pressures of seining. "We always had to take turns to make a set," Sylvia says, "so I baked rolls and took some to each boat." Once the signal blew, though, it was down to business. "We knew if they opened Anan, we would have a good year," she remarks. And good years there were. Lawrence and his crew were often able to hoist that seiner's sign of success - a broom on their mast signaling that they had netted one hundred thousand fish.

These days fishermen and regulators are on the same team, helping to make Alaska's salmon management an ongoing conservation success.

## Jim Branson, fish cop

Jim Branson was one of those whose job it was to apprehend creek and trap robbers. During the summers of the early 1940s he ran a patrol boat and, as a U.S. Fish and Wildlife enforcement officer, was empowered to arrest and bring offenders into Wrangell for trial and sentencing. He made the rounds of creeks and traps through an immense area bounded by Petersburg to the north, Kuiu Island and Port Alexander on the west, south halfway to Craig and including Anan Creek. His patrol boat, a simple, gray 36-foot craft, was underpowered for its intended job. Consequently, he was able to travel only one knot faster than the average seiner "which made a stern chase a very long one." If he caught up he took the crew to town to appear before U.S. Marshall Joel Wing who fined them. "By the time I did the paperwork they were usually back at it," Branson remarks. Being a federal enforcer earned Branson "a country full of threats" but despite those harsh words he says he got along fine during his winters in town. Running a patrol boat was a seasonal job and after two years, Branson left for the north where he eventually became the first executive director of the North Pacific Fisheries Management Council, leaving behind "a hell of a fine job in a country that was absolutely marvelous."

Alaska's political and economic impotence as a territory.

Consolidation of canneries and traps accelerated after 1930 while at the same time the number of fishermen and vessels increased. So also did the competition between fishermen and traps. Frustrated by the powerful Fish Trust's successful lobbying efforts in Washington – proposed legislation was regularly handed to fish processors for their comments – and the decline of their stocks, Alaskan fishermen took matters into their own hands. They fought the outside interests with a combination of fish piracy and fishermen's unions.

*Two men brail fish from the trap to a tender that will transport the fish to the cannery. Case & Draper photographers, 1907.*

Alaska State Library/Historical Collections/PCA 39-1000.

Fish piracy took several forms. A fish pirate ship was usually a seine boat but turn off the running lights, hang a sack over the name, muffle the motor and you have a boat ready for another kind of activity. Often these disguises were unnecessary. The fish pirate would simply make a business deal in broad daylight with the trap watchmen who lived in a shack on the trap, exchanging money for fish. Sometimes a watchman would assist by supplying a substitute lock for the one on the spiller cover. The cover prevented nets called brailers from being dipped illegally into the spiller in order to transfer fish to the seiner hold.

Another less scrupulous sort of pirate didn't even pretend to be a fisherman and had no seine aboard. He stealthily motored up to the trap under cover of darkness and took what he wanted, sometimes using the threat of weapons to complete the deal. To combat piracy the salmon packers employed a number of defenses. One or two watchmen armed with .30-30 rifles usually stayed on the trap or nearby on shore the entire season. An independently owned boat was

*Tlingit employees of the Thlingit Packing Co., 1908.*

Frank H. Nowell. University of Washington Libraries/Special Collections/NA 2192.

hired to patrol the traps and to take the watchmen off if severe weather threatened. A patrol plane might also survey the trap from above. Tender operators who hauled fish from traps to cannery were shuffled to lessen contact with pirates. In cooperation with cannery operators, the federal Bureau of Fisheries used two Navy subchasers, fully armed, to prevent fish piracy. The apex of absurdity was reached in 1951 when the Pinkerton Agency of private investigators was hired, "…to watch the patrol boats hired to watch the watchmen hired to watch the traps," noted the *Wrangell Sentinel*. Still, because of anger toward processors and the federal government alike, Alaskans developed an attitude toward fish piracy akin to that toward bootleggers during Prohibition. Apprehensions of offenders were rare and sentences usually mild.

All workers in the salmon industry looked to labor organization to help combat the power of the canneries. The first Alaskan fisherman's group was the Alaska Fishermen's Union, founded in 1902, with other organizations springing up later, particularly in the 1930s. Another was the Alaska Salmon Purse Seiners Union made up of independent seine boat operators. Each season they negotiated with the salmon industry on the price for each species of fish. When the price could not be agreed upon, seiners would sometimes resort to "tie-ups" during which they would refuse to fish. A protracted tie-up in 1938 delayed the season's opening. Shoreworkers also had their unions, all driven by the economic insecurity of an industry based on unpredictable salmon runs, seasonal employment and the abuses of an old contractor system. This system originated in the early days of the industry when Chinese workers largely manned the cannery lines. Due in part to language difficulties, a contractor made an agreement with a cannery owner to put up the fish pack for a certain price. The cannery provided transportation and housing for the workers while the contractor supplied food. The resulting system was open to much exploitation of the workers. All laborers on land and sea were dependent on the fish supply and that meant all had a considerable stake in the preservation of the salmon runs.

The White Act of 1924 had the most far-reaching effect of any fishery legislation to date. It allowed fifty percent of each salmon run to escape and spawn and also banned the granting of exclusive fishing rights to fish trap sites. Cannery operators had come to believe that they had proprietary rights to lucrative sites and would even install "dummy traps" to keep other operators out. Once again, however, the White Act appropriated no funds for enforcement so violations were rampant. A decline in salmon catches began after the peak year of 1941 with its harvest of 68 million and reached a low of 5.5 million in 1960. Alaskans' anxiety about their salmon festered. Because Congress consistently failed to allocate the resources to govern the fisheries, the abolition of fish traps became a growing and finally the primary issue in the fight for statehood. One of the first actions of the new Alaska legislature, meeting in 1959 and eager at last to assert control over its own fisheries, was to abolish fish traps, the long-endured symbol of that colonial empire, the Fish Trust.

As fisherman, legislator and fishery management leader Clem Tillion relates, once "their" fish became "our" fish, Alaskans began to actively care for the health of the resource. Beginning in 1960 recovery of the Alaskan salmon fishery began. One component was an aggressive hatchery program. However, more and more gear per boat continued to be put into service, and by the early 1970s catches had declined once more to record low levels. The alarm generated by the condition of the fishery following decades of concern about its health, led Alaskans in 1970 to approve the limited entry system. This restricts the number of fishermen working the state's waters and gives them exclusive rights to harvest the fish. Permits for each fishery were initially allocated to those fishermen who had a history of working in that fishery. From that point on, the permits could be bought and sold but no more were issued, thereby limiting the number of participants.

Control of the numbers of fishers as well as improved, well-distributed escapements in combination with near ideal environmental conditions has permitted stocks to rebound to near-record levels as noted by the Alaska Department of Fish and Game in 2003. "Eight of ten of the top escapements have occurred within the last ten years. In over one hundred years of commercial exploitation, the pink salmon harvests in Southeast Alaska are recently the highest level observed, yet the number of fish escaping the fishery to breed is also at very high levels – at the highest level since statehood, when records began."

Anan's targeted escapement ranges between .32 and .68 million per year. Seiners, gillnetters and trollers scoop up the rest. Pink salmon numbers still fluctuate, often wildly, from year to year. Not all conditions, after all, are under the control of fisheries managers, but it is to be hoped that the days of boom and bust, with potential devastation of the fishery, are a thing of the past. Nevertheless, resource managers still appear to be focusing attention on single species rather than on the sustainability of the ecosystem. Traditional Native knowledge and modern science both indicate that we ignore interconnections between species at our peril.

# CHAPTER FIVE

# Bears—Made of the Same Dust As We*

*A man belonging to the Raven clan living in a very large town had lost all of his friends, and he felt sad to think that he was left alone…so he thought he would go off into the forest.*

*While this man was traveling along in the woods the thought occurred to him to go to the bears and let the bears kill him. The village was at the mouth of a large salmon creek, so he went over to that early in the morning until he found a bear trail and lay down across the end of it. He thought that when the bears came out along this trail they would find and kill him.*

*By and by, as he lay there, he heard the bushes breaking and saw a large number of grizzly bears coming along. The largest bear led, and the tips of his hairs were white. Then the man became frightened. He did not want to die a hard death and imagined himself being torn to pieces among the bears. So, when the leading bear came up to him, he said to it, "I have come to invite you to a feast." At that the bear's fur stood straight up, and the man thought that it was all over with him, but he spoke again saying, "I have come to invite you to a feast, but, if you are going to kill me, I am willing to die. I am alone. I have lost all of my property, my children, and my wife."*

*As soon as he had said this, the leading bear turned about and whined to the bears that were following. Then he started back and the rest followed him. Afterward the man got up and walked toward his village very fast. He imagined that the biggest bear had told his people to go back because they were invited to a feast.*

*When he got home he began to clean up…When he told the other*

Ivan Simonek

*This black bear cub has been sent up a tree next to the Anan bear observatory platform. It will stay there while its mother fishes nearby. Obviously, it is just as curious about humans as they are about it.*

*John Muir, *Wilderness World of John Muir*

*people in that village, however they were all very much frightened, and said to him, "What made you do such a thing?" After that the man took off his shirt, and painted himself up, putting stripes of red across his upper arm muscles, a stripe over his heart, and another across the upper part of his chest.*

*Very early in the morning, after he had thus prepared, he stood outside of the door looking for them. Finally he saw them at the mouth of the creek, coming along with the same big bear in front. When the other village people saw them, however, they were so terrified that they shut themselves in their houses, but he stood still to receive them. Then he brought them into the house and gave them seats, placing the chief in the middle at the rear of the house and the rest around him. First he served them large trays of cranberries preserved in grease. The large bear seemed to say something to his companions, and as soon as he began to eat the rest started. They watched him and did whatever he did. The host followed that up with other kinds of food, and, after they were through, the large bear seemed to talk to him for a very long time. The man thought that he was delivering a speech, for he would look up at the smoke hole every now and then and act as though talking. When he finished he started out and the rest followed. As they went out each in turn licked the paint from their host's arm and breast.*

*The day after all this happened the smallest bear came back, as it appeared to the man, in human form, and spoke to him in Tlingit. He had been a human being who was captured and adopted by the bears. This person asked the man if he understood their chief, and he said, "No." "He was telling you," the bear replied, "that he is in the same condition as you. He has lost all of his friends. He had heard of you before he saw you. He told you to think of him when you are mourning for your lost ones."*

*It was on account of this adventure that the old people... when they gave a feast, no matter if a person were their enemy, they would invite him and become friends just as this man did to the bears, which are yet great foes to him.*

—from *The Man Who Entertained the Bears* (Swanton, p. 220-222)

Friends yet great foes. This is the paradox of bears and the source of their intense and complex interest for us. The Tlingit, as well as other native peoples, have felt an intimate connection with bears. Is it their appearance, their humanlike foot, dexterity of paw, erect stance and mask-like yet communicative features? Above all, it might be their omnivorous appetites, so similar to ours. As one writer has said, "The omnivores...are generalized eaters, and the bear is supreme among them in his taste for diversity. True to this fellowship of the open mind, its attention to potential food is broad and unrestricted, so that, for bears, the whole world is interesting. Watching them explore their environments, we recognize a consciousness somewhat like our own. We have an uncanny felling

Ivan Simonek

*Successful fishing requires concentration as this lunging black bear demonstrates. The black bear's "Roman nose" is evident in this picture.*

that beneath the fur is a man." (Shepard, p. 1) Like us, bears need to be inventive problem solvers to succeed, and succeed they have, though we have significantly diminished the numbers of our former foe.

All bears are members of *Ursidae*, a family that started out as a branch of the dog family, *Canidae*, about twenty-five million years ago. The earliest bears were forest dwellers, as is today's black bear, but were smaller than the modern black. Scientists currently believe that the ancestral brown bear migrated to North America across the Bering Land Bridge about 1.5 million years ago and that polar and brown bears then diverged from one another 300,000-400,000 years ago. On the islands north of Frederick Sound, the so-called ABC islands of Admiralty, Baranof and Chichagof, only brown bears live today, while on islands south of the sound black bears rule. Scientists at the University of Alaska have discovered that the ABC brown bears are unique. Their mitochondrial DNA is genetically different from that of all other brown bears worldwide and shows that they are more closely related to polar bears than to other brown bears. They are thought to be ancient remnants of the first brown bear population that crossed the Bering Strait, possibly three times as long ago as any other populations in Alaska. Polar bears, in fact, may prove to have arisen from these same brown bear that wandered north. Ongoing research is investigating what could turn out to be a complex story of multiple migrations and interbreeding.

Ivan Simonek

*Those claws are multipurpose tools.*

The black bear made their way to North America about 500,000 years ago via the Bering Land Bridge. While brown bears still live in Europe and Asia, the black bear is now found only in North America. Bones of both black and brown bears discovered in caves on Prince of Wales Island in central Southeast Alaska prove that both species were here well before the last glacial maximum, the time of greatest extent of the ice sheets about 20,000 years ago, and have continuously inhabited the archipelago for more than 40,000 years. A brown bear femur from On Your Knees cave is 35,365 radiocarbon years old while a black bear tibia tops that at 41,600 years old. Brown bear bones were also discovered from early post-glacial times about 14,000 years ago, soon after glaciers had receded from the island. For about two thousand years the two bear species overlapped on the islands. Today most islands in Southeast Alaska have one species or the other but not both. Bears have survived on Prince of Wales and other west coastal refugia during the last ice. Other refugia that might have served as sources of Southeast Alaska's modern bear populations were located to the south of the ice sheet and north in unglaciated parts of Alaska and the Yukon. On the now-submerged continental shelf separating the Queen Charlotte Islands (Haida Gwai) from the British Columbia mainland, another refugium might have persisted. Today about 6000 brown bears and 17,000 black bears roam Southeast Alaska.

## Black Bear (*Ursus americanus*)

Ivan Simonek

*Black bears are very comfortable in trees. See?*

Research has identified two major lineages of black bear, coastal and continental, that apparently diverged from each other in Asia as much as 1.8 million years ago. The continental lineage includes black bear from locations across North America – central Alaska, Alberta, Montana and Pennsylvania. The coastal lineage comprises all the Pacific coastal subspecies including those in Southeast Alaska north of the Queen Charlotte Islands. Further study has found, however, that both lineages occur in Southeast Alaska, an indication of the region's complex history.

Though it is called black, the species can be brown, cinnamon, blond, a rare bluish black or nearly white. In Alaska black is the norm. Their body is compact and stout with massive legs and feet. A black bear's fairly straight facial profile contrasts with the brown bear's dish-shaped profile. The lips are mobile, all the better to manipulate its varied diet. Males usually reach full growth at seven to eight years, females at five. Adult females weigh about two hundred fifty pounds while males vary between one hundred fifty and four hundred pounds.

Black bears are commonly thought to have poor eyesight but recent studies say otherwise. Evidence shows that their eyesight is comparable to ours while their night vision is superior. They can also distinguish shades of color, learning hue discrimination faster than chimpanzees and as fast as dogs. Hue discrimination serves bears well when berrying. Misconceptions about their vision may have arisen because bears do not always react to what they see. They rely more on their noses than eyes to give them the information they need. Alaskan biologist John Hechtel believes our human-biased experience often causes us to misinterpret bear behavior because we trust our vision more than our sense of smell while bears put more confidence in noses than eyes. Bear noses are truly superior organs – they can sniff out carrion a mile away – and they have other surprises in store. Despite their lumbering appearance, black bears can run up to thirty-five miles an hour and are competent swimmers. They can climb trees easily with their sharply curved claws, a legacy from their forest origins.

Ivan Simonek

*This bear, named Virginia, has been identified each year since 1997. A bear gets a name when all interpreters are reasonably sure they can pick it out as an individual. Before that they are tagged with a number corresponding to the date they first arrive at the creek. Names are chosen to reflect a bear's physical trait, Virginia's from her ear notch that one interpreter thought resembled the state of Virginia.*

The black bear is an opportunist when it comes to food and a generalist in its selection of habitat. It does well in a temperate or boreal climate as long as there is enough cover. Although they are primarily vegetarian, they will eat whatever is available but in Alaska have seasonal preferences. In spring after emerging from their dens they browse on fresh vegetation and on the occasional winter-killed animal. Bears in Southeast Alaska enjoy tender shoots and can often be seen in early summer foraging on flat grassy beaches. At Anan in spring, black bear enjoy the same herring roe on branches that the Tlingit favor. They also dig up roots of newly emerging skunk cabbage whose cathartic properties jumpstart their dormant digestive systems. In summer, when available, they feast on salmon as well as berries, ants, grubs and other vegetation. Males at times may prey on their own cubs.

Southeast black bears have a great fondness for salmon, but one study of black bear use of salmon streams on Kuiu Island, northwest of Anan, has shown that while density of bears remains high at salmon streams all season, there is substantial turnover of bears at particular streams. Most bears use the streams for less than three weeks and females frequent salmon streams in smaller numbers than males. This may be a reaction to the threat of infanticide. Females also use tidal areas less than males, preferring upstream forested areas, again perhaps because they provide protected cover for cubs.

Mating occurs in June and July. Reproduction is controlled mainly by food abundance. Those females with access to rich protein are most fertile and have healthier cubs. Females have delayed implantation meaning that the fertilized eggs are not implanted until the fall. The one to four cubs are born in late January or February while the mother is in hibernation. At birth the cubs are hairless, blind and weigh only about half a pound. They make their way to their mother's teats and begin nursing. On a diet of rich milk they gain weight rapidly, emerging from their den in May ready to explore the world. These young cubs are very vulnerable but are able to follow their mothers well. At Anan both sows and cubs come to fish. While she is at the stream, the mother

Ivan Simonek

*Mother left these three to play near the observatory while she went down to the creek. They stayed together, briefly. Like human mothers, sows find themselves trying to combine work and child care.*

# Bearly Noted

*The Forest Service interpreters, as part of their daily work, make scans every ten minutes, counting the number of people and bears. To help in identifying individual bears, a photo notebook illustrates each bear's markings, size, behavior and known background. Every adult bear is also given a name. Here are a few of the interpreters' observations.*

Achilles is a male with thin peachy coloration along his lips. He uses a large cave to eat fish and will often make many trips out for fish before leaving.

Boboli is a small adult female who sits on the knoll downstream from the observatory to eat her fish. She is very tolerant of people and will go under the observatory.

Butterfly is also an adult female with a white chest patch shaped like, what else? a butterfly.

Crow is a small female with a very long, grizzled coat. When she was first seen in 2001 her fishing ability was very poor but by the end of the season her agility and skill had improved.

Gabrielle is another female who is somewhat wary of people and will react to sounds, including camera shutters.

Gonzo (also called Speedy Gonzales) is a small male and a quick fisher not afraid to challenge larger bears for fishing spots. He will make his way slowly to the water but usually heads toward caves after catching a fish. He uses small caves that require a tight squeeze.

James Dean is a large male who will often stay around for long periods, sometimes standing in the opening of a cave and looking directly up at people on the observatory platform.

Ivan Simonek

*These twins pay careful attention as their mother, Hera, fishes. Because of her tolerance of humans, the Forest Service believes she has been coming to Anan, and probably with cubs, for years.*

Ivan Simonek

*By the end of the summer, cubs are so at ease near the observatory that they even nap on the ground.*

Juanita nursed her cubs down by the water in 2000 and has been coming to Anan since 1994.

Miss Piggy is very stocky with a thick, velvety coat late in the season. She's assertive around other bears.

Zippy may have received his name from one of several scars on his muzzle, a Z-shaped scar first noted in 1998. In 1995 he sported a "huge" wound on his left side with severe scarring in later years. He is submissive to other bears and very tolerant of people. Often, he will go under the deck to sit or eat and has even been known to sleep there in the presence of many people.

generally sends her young up a tree to protect them from larger bears. The restless cubs may whine or even run away, causing the mother to leave her fishing and go off among the brush and boulders to hunt them out. Visitors on the viewing platform are sometimes spectators to a moving drama enacted on stage across the stream as baby and mother separate, search desperately amongst brush and boulders and are, at last, reunited. With relief, everyone applauds the happy ending.

Ivan Simonek

*These adult black bears call a truce in order to take advantage of a prime fishing spot.*

The most critical time in a bear's life arrives when it becomes independent at around a year and a half. About one-quarter will lose their lives this year; another third die during the second and third years, killed in accidents and intentionally by male or female adults. After they have spent their first winter together the mother drives the cubs away to avoid the male's aggression as they seek to mate again. The newly independent cubs share their mother's range for a time, then young males will usually find their own while females share a part of their mother's range and may inherit it.

Hibernation in winter dens helps bears, both brown and black, conserve energy through the lean times. During hibernation they neither eat nor drink, yet over ninety-nine percent of black bear survive hibernation. By slowing down the metabolism of protein, excretion of waste is avoided. Because bears sometimes wake from hibernation and even groggily emerge from their dens in mild winters, some scientists instead say that bears enter into "winter dormancy" rather than true hibernation. Alaska bears den for five to seven months with sows and their cubs being the last to emerge. To prepare for hibernation, bears must bulk up on calories, often as many as twenty thousand a day. In their last few days before denning both blacks and brownies eat indigestible material to block their lower digestive track.

Bears in Southeast Alaska tend to hibernate in dens beneath large fallen trees, roots or stumps and will den from sea level to alpine areas. Good dens will be used year after year. A denning study conducted on Anan black bears found that bears selected areas protected from windstorms. These sheltered areas were more likely to harbor large, old-growth timber. In such trees, usually two hundred years old or greater, heart rot fungi create hollows favored by bears as denning sites. Although study has not shown whether the abundance of suitable dens influences bear density, bears frequently travel outside their summer home range to den, indicating that they do prefer certain types of den habitats. Thus, studies have noted that timber operations, particularly clearcutting, should be avoided in wind-protected areas where bears are likely to seek their winter dens. Once bears have located a suitable den, they will sometimes drag boughs or grass into it to make a bed or simply lay in the dirt. Often the den isn't much larger than the bear.

Scientists are looking at bear hibernation to help provide answers to some human health problems. They are looking at bear's winter activity for clues about osteoporosis. Bears do not suffer bone deterioration over the months of inactivity because they recycle calcium and phosphorus back into their bones. Shivering may help them maintain muscle tone and avoid bedsores. Physiologists studying kidney disease are looking at how bears recycle their urea, breaking it down to form protein rather than becoming toxic. Hibernating bears living off their accumulated body fat have cholesterol levels more than twice as high as humans yet show no hardening of the arteries. Hormone-like substances may control these changes in hibernating bears and suggest possible human applications.

Although bears are usually considered loners, they do socialize during the early part of their lives and in certain settings thereafter. Mother and cubs are together for the first two or three years and the siblings may be together for longer. Unexpected sights can sometimes fracture our stereotypes of bear behavior. For example, a Forest Service interpreter filed this scene in his daily report: "We do have these two male brown bears we are calling the 'brothers brown' since they are visiting the creek together. This is very unusual for adult males to do. The lighter colored bear at one time was actually catching the fish and eating part of it and bringing it up to the other bear to eat…thus came the name brothers brown." Although bears do not mate for life or roam in packs, they are constantly aware of the presence of other bears. Their home ranges overlap including species-range overlap around Anan. Through tree scratchings, excretory messages and body language, bears are continuously and subtly aware of each other's presence. At Anan the avoidance that black and brown bears normally display toward one another is set aside in a prickly truce so that both species can take advantage of the plentiful salmon.

## Brown bears (*Ursus arctos*)

Brown bears and grizzly bears used to be considered separate species but are today all classified as *Ursus arctos.* Coastal bears that feed on salmon are commonly called brown; interior bears are grizzlies. The brown bear is a very widespread species, originally native to much of the Northern Hemisphere. The species became differentiated from forest-dwelling bears during a period of glaciation when forests were replaced by tundra. In Eurasia, brown bears originally occurred from northern Africa, Spain, and Italy northward into Scandinavia and eastward through Asia to Japan. They once occupied most of Western North America, from the Great Plains to California and from Mexico to Alaska. In the lower forty-eight states the brown bear has been reduced to about two percent of its former range so that only in Alaska and western Canada are the brown bears' numbers relatively unaffected. In fact, Alaska contains over ninety-eight percent of the United States population of brown bears and over seventy percent of North America's. Geneticists have determined that there are four different clades of brown bear. A clade is a group of organisms originating from a single common ancestor and all the descendants of that ancestor. Three of the clades are found in Alaska. One is the brown bear of the ABC islands mentioned above which is closely related to polar bear and considered the oldest group of brown bears in North America. Another group is found on the Alaskan mainland and Kodiak Island, a third in extreme eastern Alaska, the Yukon and Northwest Territories and the last in southern British Columbia and a few northern states. These groupings may be a result of isolation in glacial refugia or multiple migrations from Asia.

Ivan Simonek

*This cub stays close to mother for food and safety.*

The brown bear is most easily recognized by its dish-shaped facial profile. It has a long muzzle, steeply sloped forehead and large head in proportion to its body. It is larger than the black with a prominent shoulder hump and longer, straighter claws than its black cousin. The hump and claw shape are adaptations to the open steppes and tundra where the bear developed: the claws for digging into the burrows of small mammals, the shoulder muscles for digging and also for bursts of speed while chasing caribou and moose. The size of the brown bear varies with geography. Aside from the Kodiak brown

# Play in Bears

Bears evolved from *Canids*, the dog family, quite recently. Observing bear cubs at play might, therefore, remind viewers of their family dog, especially if it was one raised with a sibling. Some researchers have hypothesized that though the two lines are now physically dissimilar, social displays such as play are quite alike. That is because, they say, social behavior changes much more slowly than does anatomy and adaptation to habitat. Examples of comparable behaviors are the submissive displays shared by both families: face-pawing, licking and nose-stabbing towards the dominant animal's mouth. Pups use all these to beg regurgitated food from their parents. Bear cubs do not obtain food in this way, but the behavior persists.

Ivan Simonek

*Tree branches make a fine jungle gym.*

Social play in black bears from cubs to sub-adults occurs among littermates, mother and offspring or in sub-adults that have separated from their mothers. Generally this play is of a type called play-fighting involving belligerent physical contact – fast running, head butting, pawing and clawing and various kinds of biting such as neck bites and muzzle seizing. Communication can mean the difference between life and death so, like a dog's "play-bow", bears use a "head-wag" signal to indicate that play is pending, not aggression.

Ivan Simonek

*Even food is fun.*

Ear posture and facial expressions are all part of cub play. As in dogs, ear position is an important part of communication in bears. Ear posture can shift from front alert to fully flattened. Crescent formations, where the ears stand out from the side of the head in a curve, or the partially flattened posture are most common during social play. Facial expressions are also important for communication in bears. Closed and open-mouthed expressions send signals ranging from relaxation to alertness. Silence is the rule when black bears play. Only when bears are seriously combative do they vocalize.

Though often solitary as adults, bears are social creatures. Play develops the bonding and communicative language that is essential in their life journey. It is rare that visitors to Anan will see cubs at play, probably in part because of the disturbing presence of humans. Instead, cubs may be intently observing the serious business of fishing or relegated to a playpen tree where they are safely out of harm's way.

Ivan Simonek

*As with dogs, this brown bear proves that a stick makes a great toy.*

Ivan Simonek

*This young brown bear will have to learn by trial and error and will have to use less desirable fishing holes this summer.*

bear, a subspecies, coastal Alaskan brown bears are the largest, with males weighing as much as 1720 pounds. More often, though, males weigh five to nine hundred pounds and females half to three-quarters of that. Coastal bears weigh more than interior bears in part because of their salmon-rich diet. Their colors range from dark brown to light blonde. Interior brown bears with silver tipped "grizzled" fur are the origin of the term "grizzly" bear. As with their black cousins, brown bears have a superior sense of smell and hearing. Similarly, their vision is rumored to be poor but this may be because they are often observed sniffing with their sensitive noses.

Brown bears are omnivores leading John Muir to comment: "To him almost everything is food except granite." (Muir, Our National Parks, p. 173) They are capable predators and meat constitutes about twenty percent of their diet. Some naturalists think that the availability of meat early in the season affects the bear's reproductive capacity. Besides salmon, they eat berries, grasses, roots of many kinds, sedges and carrion. As with black bear, brownies cycle through a seasonal round of food seeking. Upon emerging from their high altitude winter den in spring they move down to valleys and coastal estuaries for winter-killed animals, tender green shoots and small but succulent marine isopods (crustaceans) sometimes called beach lice. They then follow the greening up into the lush herb meadows of the mountains again as summer progresses. Ants, grubs, bird eggs, small mammals, roots and early berries carry them through until the salmon begin to run in July. They will gorge on these for three to four weeks then, as summer lengthens into fall, head once again up into the mountains browsing on berries, especially devil's club berries, mushrooms, fawns, mice, anything but granite. Sows with spring cubs, however, do not always frequent salmon streams due to the very real danger their cubs face from other bears. Researchers have found that these females stay further away from streams during the spawning season compared with females without young and that these females are significantly lighter in weight. Their choice to avoid salmon carries a cost that may translate into reduced female or cub survivorship.

Brown bear dens are almost always at high elevation, averaging about two thousand feet above sea level in forests, forested alpine or alpine slopes. Brown bear may excavate sizeable dens sometimes measuring up to six feet long and three feet high but generally in Southeast Alaska they use natural cavities. These may be natural cavities on steep slopes or more likely root systems of standing or dead trees. In early February two or three lightly furred newborns weighing less than a pound are born. They will stay with their mother for at least two years though in Alaska they may remain until they are three to five years old. Ninety percent of their day will be spent in foraging. Sows that bring their cubs to fish at Anan Creek usually keep to the creek mouth below the observatory. Young brown bears are generally weaned

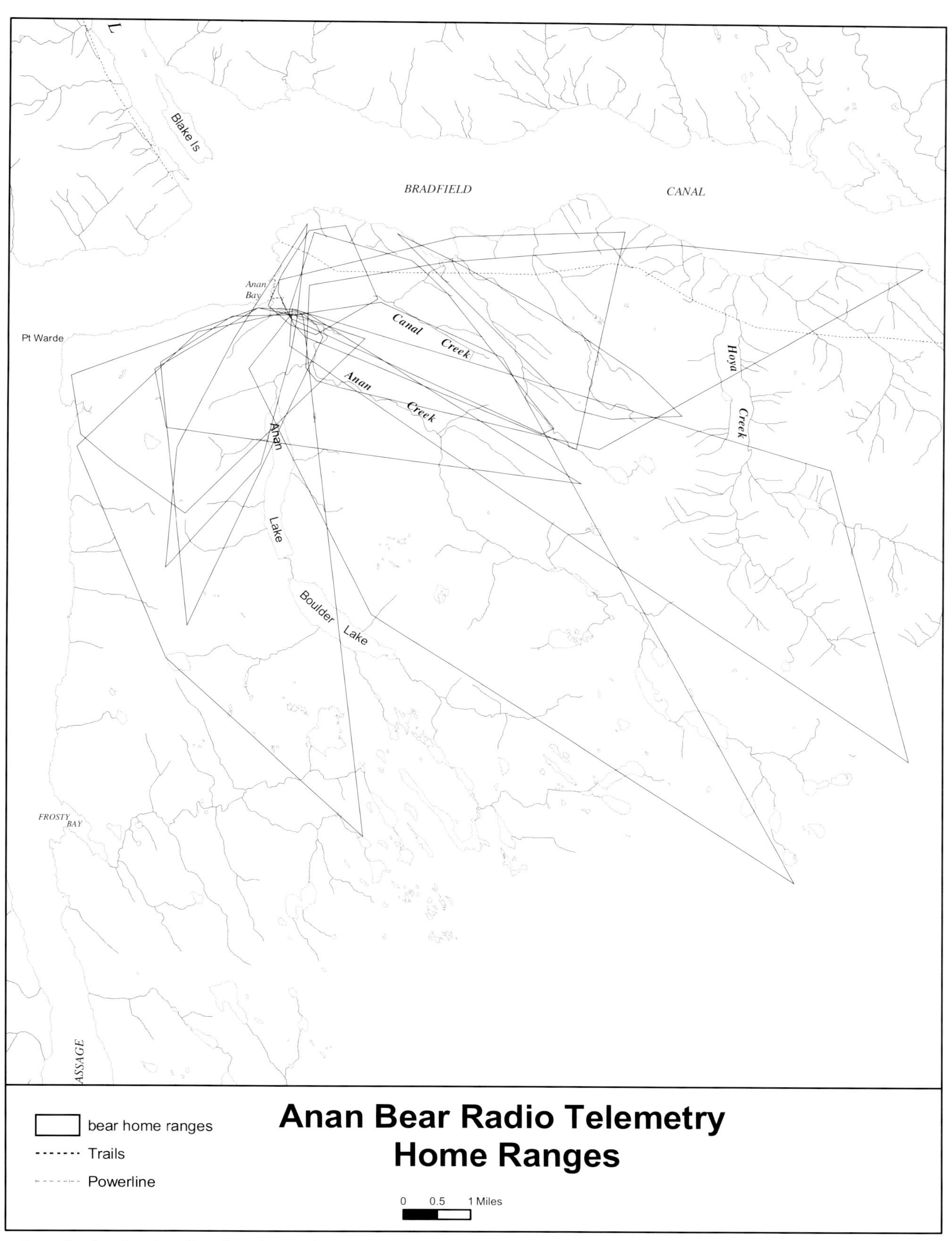


*Bears fitted with radio collars showed researchers in 1993 the extent of their home ranges including their use of Anan Creek. USDA Forest Service, Wrangell District.*

from their mothers in their third summer. Females then tend to establish home ranges near or overlapping their mother's while males wander further away.

It has been said that aggression forms the basis for social organization in all bears, but it is a finely nuanced aggression. Females avoid males except for a fleeting period of sociability. Mating takes place from May to July, peaking in early June and the pair is only together about a week. The rest of the year bears use the space of their home ranges to avoid direct encounters. The home ranges of Southeast Alaska's brown bears, those areas that meet all a bear's biological requirements, vary widely in size throughout the region. Home ranges on Admiralty Island are some of the most densely populated (with bears) in North America because of the area's abundant food resources, particularly high-quality salmon. On that island, a female's home range averages about twelve square miles and a male's forty. By contrast, that of a mainland male from the Bradfield Canal may sprawl five to six hundred square miles and will overlap those of several females. Bears spend most of their time in their home ranges that may overlap those of other bears, whom they do not attempt to exclude.

A recent three-year study of radio-collared brownies from the Bradfield Canal mainland illustrates one male's peregrinations over one year. His wanderings took him north across the Stikine River, over a snow-covered range back to the Eastern Passage, through a winter's sleep, then over to Wrangell Island and back, ending to the north again with a discarded collar near LeConte Glacier. Why all these miles? Bear technician, trapper and big game guide LaVern Beier remarks that bears are notoriously lazy creatures. The answer may partly lie in food availability but the data is still being analyzed for clues.

Beier, who has handled about eight hundred bears in the past twenty-five years, has spent the past three summers radio collaring brown bear in the Bradfield Canal for the above study as well as taking hair samples collected when bears walk through a breakaway noose located along a trail. DNA from the hair samples will be analyzed and correlated with GPS (Global Positioning System) data from the radio collars to learn more about the number of mainland bears in the area and how they use their habitat. Recent advances in genetic technology allow identification of species, sex, and individuals from DNA extracted from bear hair and scats without handling bears. Identification of individuals can be used to determine minimum population size, population trends and genetic diversity. The collars transmit a data point every half hour for a year before they fall off so researchers can see which trails and trees and fishing spots the bear prefers and where it dens.

Ivan Simonek

*This incident was provoked when the single adult came too close to the mother and cubs who only moments before had been tenderly nuzzling.*

The above study will enable the State of Alaska to more accurately determine and manage sustainable bear populations while providing for a wide range of human uses from viewing to sport hunting. Alaska is world-renowned as a brown/grizzly bear hunting area. It is the only place in the United States where they are hunted in large numbers - an average of about fifteen hundred brown/grizzly bears are harvested each year - and the trend has been increasing. Many of the hunters are nonresidents and their economic impact

*Black bears make frequent use of caves near the creek for hiding cubs, eating and napping.*

is significant to Alaska. Hunters, however, have traditionally been among the strongest advocates for bears and their habitat, providing consistent financial and political support for research and management programs.

Beier says that because of the different foods available to brown bears - highbush cranberries, mountain goats and beaver as well as different sedges - Bradfield bears use their habitat differently than island-based Admiralty bears. They also must contend with black bears. As mentioned, only one species of bear lives on most islands of Southeast Alaska but on the mainland they co-exist, black bear generally deferring to brown.

When bears do congregate, social hierarchy based mainly on size, determines who shall feed and where. Even adult males, though, give way at times to touchy sows with cubs. Researchers at Anan commented that they frequently witnessed large males scrambling up rocks to avoid confrontation with an aggressive female with cubs and (one year later) as yearlings. Body language such as laid-back ears and vocalizations that include growling, huffing, yawning and salivation are all ways that bears signal their stress. At salmon streams such as Anan, bears of both species and sexes dampen down their natural aversion to one another in order to take advantage of the salmon. The fact that there is more than enough for all and that there are many access points to the fish reduce in some measure the tension of a most stressful situation. Anan represents a unique foraging situation for bears that use mutual tolerance, accommodation and reciprocal wariness to keep costly fighting to a minimum.

Realizing that visitor numbers were rapidly increasing at Anan in the early 1990s, the Forest Service began to look for ways to first calculate, then control those numbers while also endeavoring to learn whether bear behavior, particularly feeding behavior, was being affected by these humans. In 1993 researcher Danielle Chi and others began

# Two Tlingit Bear Legends

*There are a number of versions of this story. Here is a compilation based partly on account told to Swanton in* Tlingit Myths and Texts *retelling how the brown bear became the crest of the Nanyaayih (an important Stikine Tlingit clan):*

## How the Nanyaayih Clan obtained the Brown Bear Crest

When the flood came, the Nanyaayih climbed the mountain called Sekutle'h on the south [sic] bank of the Stikine [sometimes called Cone Mountain - probably Devil's Thumb or Kate's Needle]. Two grizzly bears were also seeking refuge and in those days bears and people could talk with each other. At first the people were frightened but the bears showed by their actions that they wanted to help the people. "Don't be afraid, brothers. We'll guide you to the top of the mountain," the bears said. Finally the people reached the top of the mountain. While they waited they became hungry and killed one of the bears for food. They preserved its skin with the claws, teeth and so forth, intact. They kept it for years after the flood, and, as soon as it went to pieces, they replaced it with another, and that with still another up to the present time.

The Nanyaayih claimed the brown bear as an emblem. The signs of this trek can still be seen. On the mountain have been seen the decayed remnants of a mat and of a rope which was used to moor the raft that was used. Many songs were composed concerning it, with words such as, "Come here, you bear, the highest of all bears." This and other stories account for why the Tlingit believe that "the bear understands when we talk".

*Chief Shakes Brown Bear Canoe was adorned with the brown bear. Here it is seen in front of the Shakes clan house in Wrangell harbor probably at the end of the 1800s. The canoe is now in the Smithsonian Institution.*

Albert Niblack of the U.S Navy who was surveying the coast of Alaska for the federal government witnessed a reenactment "commemorative of the legend of the alliance of Shakes with the bear family" while visiting Wrangell in the late 1880s. Niblack said one actor played the role of the bear. The eyes, lips, ear lining and paws of a complete brown bear skin were made of copper and the jaws were capable of being worked. Chief Shakes donned a bear's ears headdress and tunic ornamented with bears' heads, then "A curtain screen in one corner being dropped, the singing of a chorus suddenly ceases, and the principal man...narrated...interrupted by frequent nods of approval by the bear when appealed to, and by the murmurs and applause of the audience." (Niblack, p. 376)

## The Bear Mother Legend

*Almost universally, indigenous peoples addressed the bear in terms of esteem, usually as an honored kinsman. The Tlingit, as well, believed it was both impolite and dangerous to refer to the bear simply as "bear" or to criticize it. Thus the bear would courteously be addressed as "my father's brother-in-law" or "one going around the woods". The following version of a widely told story teaches respect, illustrating a close human/animal connection. It dramatizes the sacrificial nature of animals voluntarily giving up their lives as a gift to feed people and also relates how the Kagwantan clan of Klukwan obtained the grizzly crest. Here is an abbreviated version.*

There once lived a chief who had many sons and an only daughter. One day the girl and her friends went berry picking and then started for home. On the way the chief's daughter stepped in bear droppings and made a rude comment about bears. A few steps farther the pack-strap of her basket broke. She fastened it but it broke again so she told her friends to keep going. It was dusk already. While she was repairing the strap she heard footsteps. She turned and saw a young man who offered her help. He picked up the basket and told her to follow him. They reached a village, but it was not the girl's home. She became the young man's wife and for a while lived contentedly with her husband's people. Soon, however, she began to see many strange things.

One night she awoke with a shock. A large grizzly bear was near her instead of her husband. It awoke with a low moan and turned into the form of the man she knew as her husband. Now she knew she was among the bear people and had been taken because she had scorned the bears. She wanted to run away but could not do it because she now had two sons with her husband.

Early the next spring her brothers went hunting in a place they had never been since their sister's disappearance. It was the place where their sister lived among the bear people. In the bears' dens, though they looked like a village to the girl, preparations began to move to summer camps. One morning the girl's husband all of a sudden started, straining his ears as if he heard something in the distance. He first looked confused, then began taking his spears down from the wall and sharpening them. All at once they heard a dog barking outside. The bear jumped up and rushed out, caught the dog and threw it in the den. It was the brothers' dog. The girl called to her husband not to resist because they were his brothers-in-law. The bear waited for the hunters to come up, then came out and gave up his life because he knew he was wrong by taking away the girl.

She came out and saw the bear lying on the snow with arrows in its side and her brothers just about to cut it. She urged them not to do it saying it was their brother-in-law. The men buried the bear and took their sister home but leaving her two sons, for they were only half human.

—Adapted from Barbeau, told by Louis and Florence Shotridge.

Besides their spiritual value, bears were of practical use to the Tlingit, although utility and reverence were inextricably interwoven. The anthropologist Frederica de Laguna writes that "The bear was the most revered animal hunted by the Tlingit, and its hunting was surrounded by religious observances." These included fasting and daily washing in cold water prior to the hunt and a special song after the kill. Because of the clan legends connecting bear and man, bear were considered kin. If a Tlingit were killed by a brown bear the men of his family or clan were obligated to form a hunting party and kill a bear. This was the same law that governed the death of a man. Usually, however, brown bear were seldom hunted in the days before guns became available in the nineteenth century, writes anthropologist George Emmons. When they were sought the preferred method was in groups with dogs using spears, snares, traps or bow and arrows. Hunters sometimes entered the den to attack the sleeping bear in the spring when their fur was prime. Snares and deadfalls were erected along trails and at the mouths of streams.

Brown bear was eaten by some Tlingit but was taboo among others. Smoked and dried, black bear was an important food source. Brown bear skins made warm bedding and ceremonial robes. The teeth, bones and sinews were used as tools and cord while the teeth were worn as amulets and in ceremonial headgear.

Bonnie Demerjian

*The Bear Up the Mountain totem on Shakes Island in Wrangell, retells the story of how the Nanyaayii clan obtained the brown bear crest.*

a three-year study, under the sponsorship of the U.S. Forest Service and the University of Utah. They looked at the effects of salmon availability and human presence on the social dynamics of Anan's bears. Rather than engage in violent displays of aggression, the study found that black bears used passive means of avoiding conflict, detouring rather than confronting each other. The most dominant bears fished where the salmon were most accessible and for longer periods of time. When brown and black bears encountered each other on the same side of the creek, black bears were more likely to give way.

Another aspect of the study found that some bears, particularly brown, fished more often at the upper falls and that they appear to be disturbed more easily by people than blacks bears. Anecdotal comments from early in the century seem to indicate that brown bear formerly were rare at Anan. A wildlife photographer did report filming a brown bear there in 1937 but most mention is of black bear. Brown bear are crepuscular, preferring the twilight hours of morning and evening to daylight. Visitor hours today are 8 a.m. to 6 p.m. and it may be that brown bears fish around this window to avoid human presence. Restricting visitor hours also means that both species of bears can utilize the lower falls and creek with less need to interact.

Black bear fished for shorter periods of time, the study learned, when groups of people at the observatory were large. A factor complicating the results of the Anan study may have been a boardwalk trail that was constructed to the bear observatory following a fall by President Ronald Reagan when he visited the site in 1992. LaVern Beier, who collared the black bears for the Chi study, says that the boardwalk acted like a fence and speculates that it changed the bears' behavior. "For the first year, they wouldn't go over it," he recalls, and believes its presence may have skewed research findings that human presence negatively affected bear feeding time. Based in part on the study, the Forest Service instituted the reservation program that limits the number of visitors at Anan to sixty-four a day with group size limits of eight.

An additional study in 1993 by the Forest Service and Alaska Department of Fish and Wildlife detailed the use of home ranges and habitat by tracking radio-collared Anan bears. The home range for female black bears averaged three and one-half square miles, for males, 13.9. The one female brown bear studied covered 11,599 acres. These bears preferred beach/estuary and riparian or riverside habitat as well as dense forest.

Why is Anan in a bear-watching class of its own? Aside from the obvious salmon runs, the caves formed by the massive boulders that line the creek are critical to these bear. Not just empty space, these dank cavities are of ecological importance. At the lower falls nineteen entrances have been counted, twenty-three at the upper falls. These serve as safe eating spots and, for females, nurseries and rest areas. Cubs can be safely secreted there while mother goes fishing. Depending on water conditions, some males were even seen fishing from caves and, at the upper falls three males were observed taking a mid-day nap in a cave. It is clear that the many functions of these rocky refuges benefit bears of all ages, sexes and species. In addition, because salmon are constricted near the base of the falls, fishing at Anan is easier for bears, who do love the easy way.

Though Anan's bears find temporary refuge in its caves and seasonal nourishment in its waters, the bears' long-term wellbeing, and that of the entire Anan watershed, need permanent protection as a wilderness area. Even then, fragmentation of bear habitat by logging roads and other development, now in the planning stages for the Bradfield Canal, may decrease the health of bears and cause increased mortality. Scientists have demonstrated that there is a correlation between cumulative miles of road construction, decreased bear density and increased mortality. Brown bear, in particular, require large unbroken expanses of land with minimal human presence. John Schoen of the Alaska Department of Fish and Game has said, "Any one development project is not a critical threat. What you have to look at is the cumulative effect. Long-term incremental change to habitat is the biggest concern...Most of the Lower 48 problems are waiting on Alaska's doorstep. We're seeing it first in Southeast; bear habitat in some areas is already being compromised by logging and associated activities such as road building." (Schoen in *Alaska's Bears*, p. 37) Black bears, too, require habitat protection. The study on Kuiu Island mentioned earlier discovered that black bear relied on small and average salmon streams rather than large waterways for sustenance. Researchers speculate that salmon may be more accessible to bears in these smaller streams.

As the majority of us become more distanced from wildness we may lose our sense of our place in the world. John Muir, who was indeed at home in the wild, reminded his readers a hundred years ago that bears had lessons to teach.

Although he speaks of the brown bears of Yosemite, his words ring true in Anan as well:

> *"Bears are made of the same dust as we, and breathe the same winds and drink the same waters. A bear's days are warmed by the same sun, his dwellings are overdomed by the same blue sky, and his life turns and ebbs with heart-pulsings like ours, and was poured from the same First Fountain. And whether he at last goes to our stingy heaven or no, he has terrestrial immortality. His life not long, not short, knows no beginning, no ending. To him life unstinted, unplanned, is above the accidents of time, and his years, markless and boundless, equal Eternity. God bless Yosemite bears!" (Muir,* Wilderness World of John Muir, *p. 313)*

We wish likewise for Anan's.

✦ CHAPTER SIX

# Long Live the Weeds and the Wilderness Yet*

Bears are the magnet at Anan Creek today and have fascinated visitors for generations, but in the past it was fishermen, trappers and hunters who were more often on the trails going about their business. In the 1930s the Forest Service took over construction and maintenance of a boardwalk trail that ran about a mile along Anan Creek. Bureau of Fisheries employees counting spawning salmon used this path that may have originally been carved out of the woods by the bears themselves. A decrepit cabin used by these men was still serviceable in the later 1930s. The woods have reclaimed it today. Another collapsed log structure above the upper falls may be the remains of an historic trapping or Fish and Wildlife Service cabin. Forest Service records from 1942 indicate that a trail maintained that year included two, now vanished, wildlife observation shelters – all ephemeral clues to Anan's past.

President Teddy Roosevelt in 1907 created the Tongass National Forest encompassing sixteen million acres of Southeast Alaska. At that time the Anan watershed came under federal jurisdiction. In the years following, the Forest Service planned and constructed about one thousand miles of trails and hundreds of cabins in the national forest for public use. One of these cabins sat on the west shore of Anan Bay in 1921. It still existed in 1941 but was later destroyed by fire. The present recreation cabin was built in 1964. Today's trail was extended beyond the lagoon in the 1930s and purposeful wildlife viewing commenced.

*Chelsea Lockabey of Wrangell snagging pinks, a family tradition. Dark areas in the stream are salmon. Courtesy Lockabey family.*

*Gerard Manley Hopkins *The Bloodless Sportsman*

## Camping with Dr. Diven

Dr. Robert Diven, pastor of Wrangell's Presbyterian Church, established a summer camp at Anan for boys and girls of the Christian Endeavor Society in 1924. For a week each summer they watched bears and birds, swam, slept in tents, fished, picked berries and explored the Bradfield Canal in skiffs towed by his larger boat.

Besides introducing Wrangell children to the wilderness Diven also had an interest in preserving Anan's bears. In the *Wrangell Sentinel* he recalled an incident with "a swagger chap" who had appeared one day while the girls' camp was in progress. "This chap…took a walk up the creek, alone. My girls were down the shore in the opposite direction. In less than an hour the gentleman returned…and rather indifferently inquired, 'What sort of hangout is this?' I briefly told him." The man then said, "Well, I'll allow this is the greatest place to see bears I've ever been in. I shot four in less than thirty minutes up there on the creek!" Diven, who cared deeply about conservation, hitched up his overalls and warned, "Young man, I'm sorry for you and all your kind; nevertheless, I'll do you a good turn for you're likely to need it. There's never been a murder committed in this camp…but evidently you don't understand the ways of the country, and we always give strangers and uninformed persons at least a running chance…If the sixteen young Amazons that belong to this camp return while you are here, and discover that you have so shamelessly shot four of the innocent bears they have been watching with so much pleasure…it will require more than one mean looking camp boss to keep them from running you into deep water at the point of your gun." Up to that point, Diven noted, the girls were making "commendable progress in their observations of both animal and bird life, as well as of the salmon." (October 28, 1932)

Diven left Wrangell in 1930 but continued to take an interest in Anan's well being and was undoubtedly pleased when, in 1937, the lower portion Anan Creek most frequented by bears was closed to black bear hunting.

*Dr. Diven's Christian Endeavor girls' camp in the summer of 1928. Wrangellite Dorothy (Voss) Ottesen attended as did Betty (Barnes) Henning, whose father Frank Barnes had been killed in a brown bear attack on the Stikine River not long before. Her mother was not pleased to learn the girls had been camping near bears with no gun at hand. Wrangell Museum.*

Hiking the Anan Creek trail to view feeding bears was popular with local residents and yachting parties alike, beginning in the first decades of the century. There were frequent lookout points for bear watching along the early trail, mostly black bear domain in those early years. In 1937 the Anan watershed was renamed the Anan Creek Black Bear Refuge and closed to black bear hunting. When the Forest Service acquired the land near the mouth of the creek in 1942 the whole became the Anan Creek Public Service Recreation Site. A bear observatory was constructed in 1965 downstream from the present location. The present one was built in 1967 and modified in 1993. There was also an observation platform at the upper falls. That has disappeared except for a few rotting planks and the trail to it is not maintained since visitors are no longer permitted there.

Today the Tongass Timber Reform Act passed by Congress designates the Anan Wildlife Viewing Area as LUD (Land Use Designation) II. LUD IIs are areas that are to be managed in perpetuity in a roadless state to retain their wildland characteristics. Unlike wilderness, limited development is permitted, such as some water and power use, mining, habitat and transportation developments under certain circumstances. Anan's 38,313 acres are thus not as fully protected as they would be under a wilderness designation. Because of its partial development (trail, cabin, observatory) Anan cannot be considered a true wilderness area as defined by the Wilderness Act of 1964: "an area where the earth and its community of life are untrammeled by man, where man himself is a visitor who does not remain." Recent changes to visitor access at Anan via a reservation system are, in part, aimed at bringing a visitor's experience at Anan closer to the solitude, remoteness and intimacy with nature that the term wilderness evokes.

Visitors to Anan Creek in earlier years were not regulated in any way. Boaters cruising through the area dropped in at will and local family groups, initially the largest users of the creek, stopped to picnic, fish and bear watch on any sunny day. Several generations of Wrangell

# Thirty-thirty Jack

In the early decades of the twentieth century yachts of the rich and famous often anchored in Anan Bay, drawn by the prospect of bear viewing. They were always met by .30-30 Jack. The *Wrangell Sentinel* reported in 1936 that, "The Anan recreational unit of the U.S. Forest Service is a popular place now for outing parties from Wrangell as well as the yacht parties cruising Alaska waters."

Jack Pratt was the fish warden at Anan Creek, employed by the Bureau of Fisheries to guard the salmon run against pirating. Such watchmen of creeks and traps were issued .30-30 rifles to add weight to their presence. Pratt was not a 9 to 5 clock-watcher, though. Anan Creek was his home, and he its most congenial host. From his cabin, set on Anan lagoon's west side, he could spot arriving boats. Then he would launch his skiff and greet the visitors before they set foot on shore. In addition to his grubby everyday clothes he had two uniforms, a sea captain's and a ranger's, and might appear in either or both in succession. No matter how he was attired, it was his enthusiasm for Anan and its bears and his stories that captivated. Jack's tall tales, baroque excesses of imagination loosely based on a lifetime of experiences in Alaska, were remembered by any who heard them.

Frontiers exert a pull of their own. Jack was one of the sourdoughs who swarmed to Alaska and the Yukon in the days of '98 looking for gold. He found it too, and, as did many others, spent it as fast as it was wrested from the ground. After one hard season's prospecting he came to town with $35,000, quickly dissipating it in dance halls. Jack had left his home in Alabama as a kid, running away to sea on a windjammer. He attended college, lived briefly in Chicago, then joined the Canadian Forces and served in France before coming to Alaska. After leaving the gold fields he worked as a deep sea diver, game hunter, "mountie", fisherman, trapper and saloonkeeper in Nome before building a cabin at Anan in 1932 to trap and earn a bit watching fish and fishermen. Here, .30-30 was as memorable an attraction as the bears. He guided visitors up the trail, widening their understanding of the site and enlivening the hike with colorful tales of his own life. A wildlife photographer remarked in 1937, "This man is the biggest__well, perhaps he just elaborates on the truth."

*Thirty-thirty Jack Pratt pictured here with two unidentified women. Courtesy Michael Nore.*

He did care for the bears, however. In the fall of the same year he settled at Anan he came into Wrangell to urge that the bears be protected in the summer. He reported that Anan was attracting yachters and that many could not resist shooting the bear. "That," Pratt remarked, "is about as much sport as shooting a friendly dog." Novelist, conservationist and friend of Teddy Roosevelt, Stewart Edward White was one of those yachters that summer. He recognized that Anan was unique and, along with some of Wrangell's leaders, worked for the bears' protection.

Hollywood stars were enamored of this larger-than-life Alaskan. Some visited Anan while filming, others were passengers on the yachts cruising Southeast Alaska. His favorite stars, Jack said, were Loretta Young and Delores Costello despite the fact that Loretta "gets the darndest yen" to clean up his notoriously unkempt cabin and once threw away his "best" winter clothes. Perhaps these stars were also attracted to someone whose expenses never totaled more than twenty dollars a month and who lived eight months of the year with only the company of his cat Bosco.

In 1939, the last year of his life, .30-30 took a trip south, his first time out of Alaska since arriving. His notoriety and exuberant personality paved the way and he was interviewed in newspapers across the country. A story in the *Chicago Tribune* and proudly reprinted in the *Wrangell Sentinel* chattered, "Thirty-thirty Jack Pratt is in town. He likes Chicago. He likes the nightclubs and the tall buildings, and all the pretty girls. But he's headed north as soon as he can get away. And when he hits Wrangell again there'll be the biggest party you ever saw…This is his first trip to "civilization" (and you can have it, he says) in thirty-eight years…'You could put all this country on a gold platter and I wouldn't touch it. Me worry the way you fellows do? There's nothing to worry about up north.'" (February 29, 1939)

He was entertained by Hollywood and went dancing one last time in San Francisco before returning to the bay he loved. Jack always intended to be buried at Anan and had prepared his own resting place. Illness, however, took him to the hospital in Wrangell at the end of the year and he never again went home. Because it was a government reserve, .30-30 could not be buried at Anan. Instead he was laid in the Wrangell cemetery without even a headstone today to pass on his story.

## Anan on the Silver Screen

Anan Creek is not only productive, it's picturesque. Hollywood thought so in 1930 when a camera crew for Radio-Keith-Orpheum arrived at the creek to shoot "atmosphere" for a feature film, *The Silver Horde*. An early masterpiece, *The Silver Horde* was released in 1930, the first year of "talkie" films. In the movie, star Joel McCrea struggles against his ruthless rival for control of the flourishing fishing industry and the affections of wealthy society gal Jean Arthur. The film was set at the George Inlet Cannery near Ketchikan with shots of that city and Anan's silvery run. Cameramen arrived at Anan after receiving word of an unprecedented run of about four hundred thousand fish that had already been counted as they passed through the weir above the first falls followed by another one hundred thousand in the lower creek. The fish were reported to be so thick in the water that they were crowding each other onto the banks of the creek in places. Five hundred twenty thousand fish was the next week's tally and no end in sight.

Paramount filmed spawning salmon at Anan Creek again in 1936 for Barrett Willoughby's novel *Spawn of the North*. The steamy flick, once called a western at sea, had a high-powered cast: Dorothy Lamour, John Barrymore, George Raft and Henry Fonda. It was set primarily in Ketchikan and included cannery scenes filmed at Waterfall Cannery on Prince of Wales Island. True to Hollywood's blasé disregard for authenticity, this is a story set in the 1890s depicting Alaskan fishermen struggling against Russian pirates who steal their traps. Never mind that the Russians gave up Alaska in 1867. The "all color" picture was filmed with the blessings of the salmon industry. It was anticipated, said a spokesperson, that, "This picture will do a great deal to impress the American public and particularly the potential tourist with the unrivaled scenery of the Territory." The film, released in 1938, won an honorary Academy Award for its special photographic and sound effects prior to those categories being added to the awards.

children ran barelegged through massed pinks while anglers built impromptu campfires to grill their catch fresh from the creek. Ruth Warfel of Wrangell remembers getting permits to snag fish at the creek in the 1940s, then taking the catch home to clean and can.

Since 2004 open access has changed, initiated by increasing visitor numbers and rising concern about their impact on the bears and vice versa. The Forest Service that manages Anan also became increasingly concerned about the quality of experience for a visitor sharing the viewing platform with twenty or more other bear watchers. Was this a wilderness experience or something else? Statistics were telling the story; the number of visitors increased from 1400 in 1991 to nearly 4000 in 2003. The type of visitor also was changing from those generally familiar with local conditions to those unused to bears. Thus in 2005 twenty-five percent of visitors were private groups and the remainder were guided or transported (dropped off at the trailhead but not accompanied to the observatory). The size of groups on the observatory platform also increased evoking comments in the visitor log complaining, "Too many people." and "I was frustrated with the number and behaviors of the visitors. Most came for an hour, were impatient, and probably were the reason no bears showed up!" Other factors were also having an effect. Anglers and bears were competing for the same fishing hole in close proximity. In several instances, bears took or attempted to steal fish from anglers. One bear took not just the fish but also the rod. These incidents took place at the creek mouth and lagoon, an important fishing area for temperamental brown bear sows and their cubs. Commercial guide use was also increasing, rising over five hundred percent between 1991 and 1995, as was use by professional photographers. Much of this increased traffic was generated by a heightened interest in nature-based tourism that was beginning to tax Anan's capacity.

Ivan Simonek

*Wrangell guide Jim Leslie leads a group of visitors up the trail to the bear observatory.*

In order to keep both bears and people safe and ensure that human use is compatible with the ecology of bears and other stream life, the Forest Service began controlling the number of visitors and group size. Their goal was to make people more predictable in time, location and behavior in order to habituate the bears. The more people are considered a neutral and predictable element by bears, the more likely they will continue to use Anan. Habituated bears bring less habituated bears close to humans and habituated sows will bring their cubs, creating generations of bears more at ease in human presence. People will be safer, too, and have a more rewarding visit.

Two Forest Service interpreters are stationed at Anan during the peak season in July and August. At an initial trailhead talk one interpreter discusses bear safety and trail and observatory etiquette. Another at the observatory provides background information on bears and other wildlife. Interpreters also collect data on bears, visitors and any bear/human interactions.

Ivan Simonek

*A very habituated black bear sow gives her cubs a lunch break at the foot of the observatory.*

During peak viewing season from July 5 through August 25 an individual pass costing $10 per person is required to visit the bear observatory. Reservations can be made up to 180 days in advance. For more information on reservations for the observatory and cabins see **Resources** at the end of this book. Fishing in Anan Creek is now discouraged during summer months to reduce the incidence of human/bear conflict. From June 15 to September 15 fishing is restricted along the trail and at the observatory. Anglers may fish in the lagoon from a boat and can cast at the trailhead. These restrictions have understandably not set well with the locals and while out-of-town visitors use Anan to the permissible limits today, many residents have ceased to visit Anan, irked, in typical Alaskan style, by unaccustomed regulation.

In addition to its federal designation as LUD II, Anan Creek is also a potential Wild and Scenic River. In 1968 Congress passed the Wild and Scenic Rivers Act that provided a means to protect rivers with "outstandingly remarkable" scenic, recreation, geologic, fish and wildlife, historic, cultural or ecological values for future generations." Only Congress can designate these rivers, and none to date have been so recognized in Southeast Alaska or the Tongass National Forest. Rivers are eligible for inclusion if they are free flowing and have at least one of the above features. Anan's salmon and wildlife viewing opportunities, says the Forest Service, make it eminently eligible for designation as a Wild River in its upper 17.5 miles and Scenic in its lower half mile. Wild River areas are defined as those that are generally inaccessible by trail and whose watersheds are unpolluted and primitive in nature – "vestiges of primitive America". Scenic Rivers are those that are still largely primitive and undeveloped but accessible in places by roads or, in Anan's case, by water and air. The Forest Service has indicated that as a potential Wild and Scenic River, Anan must be managed to protect that potential until such time as Congress approves or denies the designation. "Therefore, in addition to maintaining Anan's primitive and semi-primitive character for recreation, we are also obliged to protect

# Watching Bears: It's a tough job but somebody has to do it

*One of the two Forest Service interpreters on duty at Anan briefs visitors on trail and platform etiquette.*

*Each day in July and August two Forest Service interpreters work at either end of the Anan trail to the observatory checking passes and giving brief instruction to visitors on how to walk the trail safely and how to behave once on the viewing platform. Before the summer begins, interpreters are given extensive training in bears and how to respond to the range of bear behaviors. Bears that are habituated to humans behave differently than those who are not. The goal of interpreters is that Anan's bears become neutrally conditioned – seeing humans as neither a positive (food possessing) nor negative (cub endangering or food stealing) force. Interpreters want to ensure that bears see their visitors as "repetitious inconsequential stimuli". By repeatedly being exposed to humans who behave in an "inconsequential", non-threatening way, a bear's natural fearful response to that human will lessen over time. Visitors are told to call "Hey, bear!" to announce their presence and avoid surprising the bears, for instance, as they walk around a blind corner. Such encounters echo John Muir who wrote, "In my first interview with a Sierra bear we were frightened and embarrassed, both of us, but the bear's behavior was better than mine." Interpreters aim to forestall those embarrassing moments.*

*Sometimes a bear will engage in some undesirable, but non-aggressive activity. Should a bear begin sniffing curiously through the rails of the viewing platform, the interpreter, to deter or alter the bear's behavior, might use aversive conditioning such as bear spray or a loud noise. Occasionally, says Dee Galla of the Forest Service, bears, particularly three to five year old brown bears away from their mothers for the first time, will make a game of gradually moving closer to people. "They are testing you like they do other bears," she says of these "hooligan" youth. Non-submissive posturing, grouping together and "looking big" without backing up can all help change a bear's mind.*

*The following are excerpts from interpreters' weekly shift reports illustrating moments of tension and satisfaction during their days of bear and people watching:*

*Gates at the observing platform keep the very curious at bay. Interpreters may also step in to enforce bear compliance.*

There were no bear incidents during the day this week…Late evening is another story however. The interpreters have been recommending that people not visit after 19:30 due to increased brown bear activity. [Night viewing is no longer allowed.] Sunday night, the cabin user was on the observation deck along at about 20:00. He reported four brown bears approaching the observation deck. One bear put its front legs on the gate and then when it started to lift a back leg, he started shouting at the bear. This was after he checked to see if he could get on to the observatory roof. It eventually left along with the others…Monday night the bears were again frisky. They were chasing each other across the lagoon mud flats (low tide) and fighting in the woods.

There are approximately five different brown bears that have been frequenting the area. One of the brown bears is a large, handsome sow that apparently has spent numerous hours on the Nautilus machine.

The first salmon were sighted on June 25, and by the end of the week a good number were in the creek, enough for the bears to have success fishing… the eagles seem to like the lower water, for they can fish in places normally that are too deep.

Luna [one of the brown bears] did have an opportunity to kill a two year old black bear, as she had it cornered by the lower end of the fish pass and also within her reach, but she chose to leave it alone after a minute or two and to fish instead. Next week—Luna has been the main attraction at the lower falls.

Luna has chased away many different black bears and seems to be claiming the good fishing for herself...We are thinking of changing the name Anan to "Luna's World". Next week--Welcome to the Queen Luna Fish Pig show. When going to Anan these days, you are very likely to see this bear, either on the trail or at the observatory, as she is quite often at the observatory for several hours straight, catching as many as fifty salmon during that time.

The river otters are putting up a good show…Up to six otters have been seen fishing and playing near the observatory at a time. Seeing one wrestle and land a salmon is quite the spectacle.

Boboli, a young bear we're very familiar with who habitually uses the area below the platform (with or without visitors on it) as a cave, displayed intensely curious behavior toward a young boy on the platform…the bear placed both paws up on the deck and stuck his muzzle between the slats…I stomped up from the lower portion of the platform, said "Hey bear," continued to approach, yelled "Hey bear!" then stepped between the child and the bear and clapped my hands. The bear did not respond to my actions. I sprayed the bear with a one-second blast of red-pepper spray at a range of 6-12 inches. The bear ran out of the area.

Try to relate to the bears as the bears would relate to each other. By observing the way the bears interact, we as humans can become more aware of how we might best be able to convey messages which are understandable to the bears…by watching how bears react to encounters with each other, we can emulate some of the conflict avoidance behavior which they utilize.

Ivan Simonek

*There are plenty of opportunities for close-ups on the viewing platform.*

The fish were just starting to show up…There was a family from Mexico that even though there were no bears and not too many fish they went away happy because of the beauty of the land.

There was this one young brown bear that was about 2 ½ years old that seemed to be a little out of place. We all decided that this must be his first time on his own. It was quite comical to watch him try to catch the pink salmon. It took him quite a long time to figure out exactly how he needed to fish.

Ivan Simonek

*These two black bear adults uneasily share a space as they fish.*

Anan had a high number of visitors this week…There seem to be a lot of people visiting from all over the world… There were six men staying at the Recreation cabin for a week from New York, Las Vegas, Paris and Florida. They were a pleasure to have around and I enjoyed watching them experience their visit to Anan. They thought Anan was so wonderful they are already planning on a trip back next summer.

There have been so many new bears every day…Each bear is so different in so many ways and it is very interesting to see the visitors try to ID the bears… We do have a cub finally…The mother is obviously a new mom since she tends to lose her cub and an hour later she has to go search the treetops for her child. I really hope she gets the hang of the motherhood routine. I would hate to witness her losing her cub.

The spawned out fish are being washed out to sea and the bears are grabbing them as they hit the shore to eat whatever is left on the fish. One bear we are calling Aqua-bear actually swam out into the lagoon to get a floater. Aqua-bear is a brown bear that is 2 ½ years old and is bored of fishing and has been taking a daily swim around the lagoon.

There are lots of seals showing up outside the lagoon, which can mean only one thing…the fish are not too far away…Other than the seals there are a lot of mink being seen running around searching for food. We have even had the chance to watch them dive for crab. I can't believe the size of the crab they can catch!

The lack of bears has presented a challenging opportunity for the interpreters. Trying to show people how diverse the ecosystem around Anan Creek is, even without bears, often difficult in the face of unmet expectations of large, toothed, majestic bruins. The river otters help, as do the harbor seals, the mergansers, the eagles, the salmon, etc. It is quite an amazing place, and the more time one spends there, the more one sees.

# Tips for Successful Photographs at Anan

*Close contact with bears makes Anan a special place and everyone wants their own bear photos to remember the day. However, challenging lighting and bears that are sometimes too close or too active can lead to blurry pictures. Following some simple suggestions by Wrangell photographer Ivan Simonek, who contributed most of this book's color photographs, can help produce memorable pictures.*

**What to do first**

As soon as you get to the observatory, sign up for the blind. The blind is an enclosure situated closer to the creek than the observatory platform but connected with it by a camouflaged staircase. Bears will be fishing right below it only a few feet from your camera. The Forest Service permits only five people at a time in the blind with a time limit of thirty minutes per person a day or less, depending on how many visitors are on the platform. The blind is your best location and no telephoto lens is needed. You can sit on a stool and support your camera on a windowsill. While on the observatory platform you will also have plenty of photo opportunities.

**Equipment**

While sophisticated equipment will give you a better chance of success, what counts most is correct tactics. One-time use cameras are not recommended. Zoom lenses come in handy and, since you may be as close as six feet from a bear, a wide-angle lens is useful. If you can cover a range of 28 to 400 mm, you will be ready for anything. Image stabilization can also be a useful feature.

**Low light situations**

Flash photography at or around the observatory is not allowed so you will be dealing with slow shutter speed and blur due to camera and subject movement. While a tripod eliminates camera movement, I don't recommend one because on the observatory platform you will be constantly changing your position since bears arrive from all directions. A monopod is better but what works best is bracing your body, arms or the camera against a solid object. If your flash can't be turned off, place a piece of tape (available at the site) over the flash. Photographers using film cameras should use ISO400 film at a minimum. ISO800 or ISO1600 is better. Those using high end digital SLRs can increase the camera sensitivity to a similar setting. The general digital public can't match such settings without inviting too much digital noise into shadow and black bear fur so I don't recommend setting your consumer digital camera any higher than ISO200 or, at most, ISO400.

While a higher sensitivity setting eliminates both primary causes of blur for those visitors with more sophisticated equipment, most will have to use that bracing tactic to eliminate camera movement. But what about subject movement? The moment arrives when a bear successfully grabs a fish and turns to run with it, a moment that slow shutter speed just can't capture. Keep your cool and time your shots for the brief pause that happens when a bear turns, climbs over a boulder, etc. Wait for anything that slows down a bear's movement. It helps to anticipate the bear's path and action. Pre-focus by pressing your shutter release button halfway down and holding it until the right moment.

If the fishing conditions are favorable and the bear experienced, you may have the bear in sight for only ten seconds. This applies to black bear, the ones you will see most frequently. Generally, black bear have their fish "to go" and eat it out of sight. Brown bear, on the other hand, move slowly and deliberately, eating without hurry in plain sight. Brown bear also photograph much better because they do not blend into shadows and dark background while their facial features stand out.

Ivan Simonek

*This scene of a black bear peering from a dark-toned boulder cave illustrates the challenges of photographing bears at Anan Creek.*

A close up of a black bear also has the tendency to fool your camera's lightmeter so you will want to compensate the exposure. A black bear against a dark background is a scene with very low contrast so if your digital camera has settings for higher contrast, use them. The exception is a black bear standing in white water. Such a scene requires the opposite – lower the contrast to a minimum.

**Focusing**

Automatic focusing is usually based on contrast so some cameras may have a hard time properly focusing on a black bear. If you pre-focus on a gray boulder at the same distance as the bear and lock your focus and exposure by holding your shutter release halfway down, that should solve the problem.

Be sure to bring plenty of film or memory cards. The question will not be *if* you see any bears, but how many you will see. Good luck and see you there!

*—Ivan Simonek*

## Anan Cabins

Sometimes an hour or two at Anan isn't enough. If you want leisure to absorb more of Anan's riches in greater seclusion, consider renting a Forest Service cabin for a night or two. There are two cabins: the Anan Bay cabin overlooking the bay is reached by boat or float plane, Anan Lake cabin is on the south end of Anan Lake and is accessible only by float plane.

The Anan Bay cabin is an A-frame with loft that sleeps seven. Anan Bay cabin users are required to have permits to visit the bear observatory, as are all visitors to Anan from July 5 to August 25. Four permits will be held for purchase for each day the cabin is reserved. A half-mile trail connects the cabin with the Anan Creek trailhead. A further half-mile trail leads to the observatory. As mentioned in the text, fishing is restricted in summer months along the trail and at the observatory but allowed from a boat in the lagoon and from the trailhead. See **Resources** for information on cabin reservations and permits. Fishing licenses are required for both resident and non-resident anglers.

Bonnie Demerjian

*Anan Bay Cabin.*

The Anan Lake cabin is more remote and accessible only by floatplane. Although a trail from the shore to Anan Lake cabin appears on topographic maps it is not maintained and the lower portion passes through an area now closed to the public in summer months. A skiff and oars, but no life vests, are provided. A rough trail about .3 miles from the cabin leads to Boulder Lake. In both lakes fishermen can find steelhead trout (April and May), rainbow and cutthroat trout (June - September) and silver salmon (August - October). In Anan Creek there are steelhead (April - June), cutthroat (May - September), Dolly Varden and pink salmon (July - August) and silver salmon (August - October). Fishermen have been quite successful in these waters. Al Oglend, a Forest Service technician who helped build and maintain the Anan Lake cabin, also maintained a long tradition of embellishment among fishermen when he remarked, after throwing a line in Boulder Lake, that the cutthroat were so thick he had to hide behind a tree to bait his hook.

*Anan Lake Cabin. Courtesy of USDA Forest Service, Wrangell District*

Anan's remarkable bear and other wildlife resources." (Final SEIS, Tongass Land Management Plan Revision, Vol. I, p. 3-182--3-187)

Timber cutting and the roadbuilding that accompanies it may someday occur not far from Anan Creek. The Canal Hoya sale was approved by the Forest Service in 1999 and is expected to yield fourteen million board feet of timber from six hundred sixty acres of land. Canal Creek is located about three miles east of Anan Creek and Hoya Creek approximately eight miles beyond that. The area to be cut is sandwiched between Anan and the brown bear habitat of the Eagle River drainage further east. Prior to decision-making, the home range study of radio-collared Anan bears discussed earlier showed that they do use the western portion of the proposed cut area.

During the public comment period on the Tongass management plan, many wildlife biologists familiar with Anan raised concerns about the cut and its ecological risks. Roadbuilding opens up land to ATVs and hunting. Bears already acclimated to humans, such as are Anan's, may be at more risk. Many criticized the limited scale of the home range study - only 14 black bears and one brown were collared – and asked that it be broadened. Another concern of biologists is that the southern portion of the cut area is also mountain goat range. In order to address these issues, the

## A Walk in the Woods

*Despite its dense black coat and considerable bulk, a black bear can silently appear as if conjured out of the underbrush itself.*

There's something that's different about walking in the woods with bears. Most of us have become so removed from the experience of walking through a place inhabited by truly "wild" life that we feel slightly out of control, as indeed we are when we push through greenery that just might conceal something larger than ourselves. And the landscape in the wild is not well groomed. At best it is hacked away, here at Anan, primarily to keep at a minimum those surprise encounters between man and beast that are so mutually disturbing. Wild animals are not remote-controlled. This is their turf and has been for much longer than we can imagine. The job of the Forest Service interpreters is to make us as ordinary as a stump. Not a threat, not overly remarkable, just another part of the whole. That's certainly not the way we are accustomed to feeling but it is most appropriate here at Anan and imperative in the wider world.

We tread the soggy path to the observatory with all senses alert, if we have any wit about us, if we're not completely dulled by modern life. The lush rainforest should be enough to wake us up, but forest mixed with bears should rouse every nerve. How instructive it is to walk amongst animals that could do us considerable damage. This is what wilderness is really about. Take away the bears and you have a park, a greenbelt. Many Alaskans assert, in fact, that they would rather roam the woods with bears than take their chances with two-footed antagonists on a dim urban street.

We might be inclined to say that we're sharing the woods with bears but, equally, they are sharing it with us. Over and over in accounts from indigenous people, wilderness guides, biologists, all of whom encounter bears frequently, the bear is rarely the ravening, drooling monster of popular legend. "Bear? They always walk around, they never bother us. ...We just say 'go on, go on,' and he understands it," recounts one Tlingit elder about her long-ago fish camp days. Our new ecological understanding of the earth places all life and all elements in relationship with each other. We need to remember again and again that we are a part of an ensemble. Walking with bears can be instrumental in reminding us, forgetful as we are, of *our* place in the woods.

Canal Hoya final decision plans no roads in the Canal Creek drainage adjacent to Anan. Helicopter yarding or ingathering of logs will be used to harvest timber north of a power line that runs from east to west through the area. Six miles of permanent road will be built beneath the power line in the Hoya Creek drainage to harvest timber south of the line meaning that roads will lie six and one-half miles from the Anan observatory. The Forest Service plan defers decisions about further cutting and road building to access more timber in the Canal Creek drainage closer to Anan. In 2005 a federal judge in Alaska blocked nine planned timber sales in the Tongass National Forest until managers complete a new Tongass Land Management Plan (TLMP) or until 2008, whichever is earlier, using new estimates of wood product demand. Estimates of timber demand in the TLMP were ruled to be inaccurate. Canal Hoya, Anan's neighbor, is one of the proposed projects whose fate hangs in the balance and may be decided in 2007.

As we have seen, Anan's bears are indifferent to management decisions, environmental impact studies and timber sale plans. Eagles, seals and salmon, too, have their own agendas. They are not, however, immune from their effects. We are trustees of the land and water in all their abundance and are compelled to protect them for as far into the future as we can see and further yet. Throughout this book, I have tried to capture the sense of deep time as embodied in a specific place – Anan Creek. Deep time is the concept of geologic time first recognized by James Hutton, a Scottish geologist living in the 1700s and considered by many to be the father of modern geology. Contrary to the conventional wisdom of his time, he believed that Earth is very old. So old, in fact is it that, in Hutton's view "We find no vestige of a beginning, no prospect of an end". What we know about the history of earth has been clarified since his day but he did help stretch our vision backward into unfathomable time. Looking into the future is another story. Are we willing to acquiesce to warnings that species will disappear in twenty years, fifty, one hundred? Salmon populations in Alaska are as healthy today as those prior to white settlement, thanks in part to good management and to prime environmental conditions. Alaskan bears, also, are thriving. But what if the environment changes - seas warm, streams silt up -and development practices fracture habitat? Deep time is not only retroactive but extends into the future. Imagination fails us as we try to see what lies ahead. All the more reason to harness science and religion, data and myth as we strive to make wise choices about our world.

The words of Aldo Leopold, considered the father of wildlife conservation in America are a fitting conclusion. He said:

> *Conservation is a state of harmony between men and land. By land is meant all of the things on, over, or in the earth. Harmony with land is like harmony with a friend; you cannot cherish his right hand and chop off his left. That is to say, you cannot love game and hate predators; you cannot conserve the waters and waste the ranges; you cannot build the forest and mine the farm. The land is one organism. Its parts, like our own parts, compete with each other and co-operate with each other. The competitions are as much a part of the inner workings as the co-operations. You can regulate them—cautiously—but not abolish them.*
>
> *The outstanding scientific discovery of the twentieth century is not television, or radio, but rather the complexity of the land organism. Only those who know the most about it can appreciate how little we know about it. The last word in ignorance is the man who says of an animal or plant: "What good is it?" If the land mechanism as a whole is good, then every part is good, whether we understand it or not. If the biota, in the course of aeons, has built something we like but do not understand, then who but a fool would discard seemingly useless parts? To keep every cog and wheel is the first precaution of intelligent tinkering. (Leopold, p. 145-146)*

Hunting and gathering people the world over understand that all sustenance is a gift whose value exists not only for those in the present, but also for those who are yet to be. How we watch over our heritage today, including Anan, that stream of living water, will shape our world tomorrow.

# Resources

## Suggested Reading

BEARS

*Alaska's Bears.* Alaska Geographic Society

*Bear Facts: the essentials of traveling in Bear Country*
http://www.fs.fed.us/r10/tongass/forest_facts/safety/bearfacts.htm

Brown bear info website – lots of links to bear information
http://ebiomedia.com/teach/BearLinks.html

*The Grizzly Bear.* Thomas McNamee

*Grizzly Country.* Andy Russell

*The Sacred Paw.* Paul Shepard and Barry Sanders

COMMERCIAL FISHING AND CANNERIES

*Alaska's Salmon Fisheries.* Jim Reardon, ed.

*The Silver Years.* Lawrence Freeburn, ed.

GENERAL NATURAL HISTORY

*Southeast Alaska's Natural World.*
Robert H. Armstrong and Marge Hermans

GEOLOGY

*Geology of Southeast Alaska.* Stowell, Harold H.

PLANTS

*Native Plants of Southeast Alaska.* Judy K. Hall

SALMON

*Reaching Home: Pacific Salmon, Pacific People.* Natalie Forbes

*Salmon Without Rivers: A History Of The Pacific Salmon Crisis.* James A. Lichatowich

TLINGIT INDIANS

*Indian Fishing – Early Methods on the Northwest Coast.* Hilary Stewart

*Monuments in Cedar.* Edward L. Keithahn

*The Social Economy of the Tlingit Indians.* Kalervo Oberg

*The Social Structure and Social Life of the Tlingit in Alaska.* Ronald L. Olson

*The Tlingit Indians.* George Thornton Emmons

*The Tlingit Indians.* Aurel Krause

*Tlingit Myths and Texts.* John Swanton

## How to get to Anan Creek

To make Anan Wildlife Observatory reservations or to rent a cabin at Anan Bay or Anan Lake go online to the National Recreation Reservation Service. You can reach this nationwide service:

by phone at 1-877-444-6777 (toll free call), international 518-885-3639

by internet at http://www.fs.fed.us/r10/tongass/recreation/wildlife_viewing/ananobservatory.shtml

The cost for the Anan Bay cabin varies by season. The non-peak season charge, from 10/1 - 4/30, is $25 per night. During the peak season, from 5/1 –9/31, the charge is $35/night. The Anan Lake cabin is $25 year round.

To get to Anan Creek Wildlife Viewing Site:
http://www.fs.fed.us/r10/ro/naturewatch/southeast/anan/anan_getthere.htm

## For outfitters, transporters and guides contact:

Wrangell Convention and Visitor's Bureau
1-800-367-9745 or

USDA Forest Service, Wrangell Ranger District
907-874-2323

Wrangell Chamber of Commerce
www.wrangellchamber.org

Ketchikan Chamber of Commerce
www.ketchikanchamber.com

Petersburg Chamber of Commerce
www.petersburg.org

A list of authorized guides and transporters can also be found at http://www.fs.fed.us/r10/tongass/districts/wrangell/anan_guides_transporters.shtml

# Bibliography

Ager, Tom. Ecosystem and climate history of Alaska, Southeast Region. USGA Earth Surface Dynamics Program, 2004. http://esp.cr.usgs.gov/research/alaska/index.html

Alaska's bears. Alaska Geographic Society. Anchorage, AK, Vol. 20, No. 4, 1993.

Alaska's salmon fisheries. ed. Jim Reardon. Anchorage, AK: Alaska Geographic Society, 1983.

Anan Creek History. Unpublished from Wrangell District, USDA Forest Service files. n.d.

Anan Management Standards, Environmental Assessment Stikine Area. USDA Forest Service, Alaska Region. R1—MB-317, 1996.

Andrews, Clarence L. The Story of Alaska. Caldwell, ID: Caxton Printers, 1938.

Barbeau, Marius. Totem poles: according to crests and topics, Vol. 1. Hull, QU: Canadian Museum of Civilization, 1990.

Barrick, Lowell. Factors affecting the choice of a fishway design for Anan Falls. Juneau, AK: University of Alaska Southeast, 1977.

Ben-David, Merav, Titus, Kimberly and Beier, LaVern R. "Consumption of salmon by Alaskan brown bears: a trade-off between nutritional requirements and the risk of infanticide?" Oecologia, Vol. 138, No. 3, February, 2004.

Bonar, S.A., G.B. Pauley, and G.L. Thomas. 1989. Species profiles: life
histories and environmental requirements of coastal fishes and invertebrates (Pacific Northwest)--pink salmon. U.S. Fish Wildlife. Service. Biol. Rep. 82 (11.88). U.S. Army Corps of Engineers, TR EL-82-4. 18 PP.

Browning, Robert J. Fisheries of the North Pacific, Anchorage, AK: Alaska Northwest Publishing Co., 1974.

Canal Hoya Timber Sale: Final Environmental Impact Statement. Petersburg, AK: United States Department of Agriculture, Forest Service, Alaska Region, 1998.

Capra, Fritjof. The web of life. New York: Anchor Books, 1996.

Chi, Danielle K. The effects of salmon availability, social dynamics, and people on black bear (*Ursa americanus*) fishing behavior on an Alaska salmon stream. Logan, UT: Utah State University, 1999.

Carrara, P.E., Ager, T.A., Baichtal, J.F., and VanSistine, D. Paco , Map of glacial limits and possible refugia in the southern Alexander Archipelago, Alaska, during the late Wisconsin glaciation: U.S. Geological Survey Miscellaneous Field Studies Map MF-2424, U.S. Geological Survey, Denver, Colorado, 2003.

Christensen, Bob and Van Dyke, Cheryl. Brown bear (*Ursus arctos*) habitat and signs of use: Berner's Bay, Alaska site survey – June 15-19, 2003. Juneau, AK: Southeast Alaska Wilderness Exploration, Analysis & Discovery.U.S. Department of the Interior, U.S. Geological Survey, 2003.

Cobb, John N. Pacific salmon fisheries. Washington, D.C.: U.S. Government Printing Office, 1930.

Colt, Steve. Salmon fish traps in Alaska: an economic history perspective. Anchorage, AK: Institute of Social and Economic Research, University of Alaska Anchorage, 1998-2004. http://www.alaskool.org/projects/traditionalife/fishtrap/FISHTRAP.htm

Conservation of the fisheries of Alaska. Juneau, AK [?]: Alaska Territorial Fish Commission, 1923.

de Laguna, Frederica, "Tlingit", in Handbook of North American Indians, Vol. 7, Northwest Coast. Washington, DC: Smithsonian Institution, 1990.

DeBruler, Ralph. "Here comes a fish pirate!", The Alaska Sportsman, September, 1941.

Eberhart, V.A. "The Fish Pirates", The Alaska Sportsman, Vol. 7, No. 9, May 1951.

Emmons, George Thornton. The Tlingit Indians. Vancouver, BC: Douglas & McIntyre, 1991.

Tongass Land Management Plan, Final Environmental Impact Statement, Vol 1, Appendices A, B, D, E. Juneau, AK: USDA Forest Service R10-MB-481a, 2003.

Forbes, Natalie. Reaching home: Pacific salmon, Pacific people. Anchorage, AK: Alaska Northwest Books, 1994.

Gay, Joel. Commercial fishing in Alaska. Anchorage, AK: Alaska Geographic Society, Vol 24, No. 3, 1997.

Gende, Scott M. and Thomas P. Quinn. "The fish and the forest". Scientific American.com, August, 2006. http://sciam.com/articleID=000C6B0F-B1A0-14C0-B04F83414B7F0000.

Goldschmidt, Walter R. and T. H. Haas. Haa Aani Our Land: Tlingit and Haida land rights and use. Seattle, WA: University of Washington Press and Juneau, AK: Sealaska Heritage Foundation, 1998.)
(Reprint of Possessory Rights of the Natives of Southeast Alaska. Washington, DC: U.S. Government Printing Office, 1946.)

Gregory, Homer E. and Barnes, Kathleen. North Pacific Fisheries, Studies in the Pacific No. 3, San Francisco, CA: American Council Institute of Pacific Relations, 1939.

A guide to the Alaska Packers Association records and to the Alaska Packers Library. Juneau, AK: Alaska Historical Library, n.d.

Haida Gwaii: Human history and environment from the time of Loon to the time of the Iron People..Gedje, Daryl W. and Rolf W. Mathewes, ed. Vancouver, BC: UBS Press, 2005.

Henry, J.D. and S. M. Herrero. "Social play in the American black bear: its similarity to *Canid* social play and an examination of its identifying characteristics." Integrative and Comparative Biology, Vol. 14, 1974.

Holfjeld, Johannes. "My six weeks on a salmon trap", Alaska Life, Vo. 8, No. 10, October 1945.

Keithahn, Edward L. Monuments in cedar. New York, NY: Bonanza Books, 1963.

Kingsbury, Alan. "Pink salmon," Alaska Wildlife Notebook. Alaska Department of Fish and Game, 1994. http://www.adfg.state.ak.us/pubs/notebook/fish/pink.php

Krause, Aurel. The Tlingit Indians. Seattle, WA: University of Washington Press, 1956.

Langdon, Steve J. "Tidal pulse fishing: selective traditional Tlingit salmon fishing techniques on the West Coast of the Prince of Wales Archipelago," Traditional ecological knowledge and natural resource management, Charles R. Menzies, ed. Lincoln, NB: University of Nebraska Press, 2006.

Leopold, Aldo. Round River, Oxford University Press, New York, 1993.

Liberate Alaska from the fish trap, compiled by Ketchikan and Cordova fishermen. Ketchikan, AK, 1949.

McCormick, Samuel J. Alaska Journal of Samuel James McCormick June 6, 1937-September 18, 1937. Fairhaven, MA: Vining Press, 2003.

McNamee, Thomas. The grizzly bear. New York: Penguin, 1984.

Miller, Mike. "The story of statehood", Alaska Airlines magazine, October 1984.

Miller, Polly and Gordon. Lost heritage of Alaska. New York, NY: Bonanza Books, 1967.

Moore, Kathlee and Jonathan Moore. "The gift of salmon", Discover, Vol. 24, No. 5, May 2003, pp. 44-9.

Moser, Jefferson F. The salmon and salmon fisheries of Alaska. Report of the Operations of the United States Fish Commission Steamer "*Albatross*" in 1900 and 1901.Washington, CD: U.S. Government Printing Office, 1902.

Muir, John. Our National Parks. San Francisco, CA: Sierra Club Books, 1991.

Newton, Richard and Madonna Moss. The subsistence lifeway of the Tlingit people: excerpts of oral interviews. United States Department of Agriculture, Forest Service, Alaska Region report No.179, n.d.

Niblack, Albert P. The Coast Indians of Southern Alaska and Northern British Columbia; based on the collections in the U.S. National Museum, and on the personal observation of the writer in connection with the survey of Alaska in the seasons of 1885, 1886, and 1887. New York, NY: 1888. [Johnson Reprint Corp., 1970]

Nowacki, Gregory et al. Geographical Subsections of Southeast Alaska and neighboring areas of Canada. Washington, DC: United States Department of Agriculture, U.S. Forest Service, Alaska Region, 2001.

Oberg, Kalervo. The social economy of the Tlingit Indians. Seattle, WA: University of Washington Press, 1973.

Olson, Ronald L. Social structure and social life of the Tlingit in Alaska. Berkeley: CA: University of California Press, 1967.

Olson, Wallace M. The Tlingit: an introduction to their culture and history. Auke Bay, AK: Heritage Research, 1991.

Orth, Donald J. Dictionary of Alaska Place Names. Washington, DC: U.S. Government Printing Office, 1967.

Pacific salmon life histories. Groot,, Cornelius and Margolis, L. ed. Vancouver, BC: University of British Columbia Press, 1991.

Paul, William L. Sr. "Fish and fish traps: the Indian viewpoint", Alaska Life, Vol. 11, No. 7, July 1948.

Peacock, Elizabeth. "Quantification of black bear use of salmon streams," Population, genetic and behavioral studies of black bear Ursus americanus in Southeast Alaska. Reno, NV: University of Nevada, 2004.

Peck, Cyrus. E. The tides people: Tlingit Indian of Southeast Alaska. Juneau, AK: City & Borough of Juneau School District, 1975.

Porter, B. "Unit 1 brown bear management report", Brown bear management report of survey and inventory activities 1 July 2002-30 June 2004", C. Brown, editor Juneau, AK: Alaska Department of Fish and Game, 2005.

Price, Robert E. The Great Father in Alaska: the case of the Tlingit and Haida salmon fishery. Douglas, AK: First Street Press, 1990.

Report of the Governor of Alaska to the Secretary of the Interior, 1899. Washington, DC: U.S. Government Printing Office, 1900.

Rogers, George W. Alaska in transition: The Southeast. Baltimore, MD: Johns Hopkins Press, 1960.

Rogers, Lynn L. "Bear center conducts hibernation study – fall 1999", http://www.bear.org/Black/BlackBearResearch/Bear_Center_Conducts_Hibernation_Study.html

Roppel, Patricia. Alaska salmon hatcheries, 1891-1959. Alaska Historical Commission Studies in History No. 20. Portland, OR: National Marine Fisheries Service, 1982.

Roppel, Patricia. "In search of the Point Warde Cannery", Alaskan Southeaster, January, 2003.

Roppel, Patricia. "Pershing asked for canned salmon", Southeastern Log, September, 1988.

Roppel, Patricia. Southeast Alaska, a pictorial history. Norfolk, VA: Donning Co., 1983.

Rozell, Ned. "The brown bear, father of the polar bear?" Alaska Science Forum, #1314. Fairbanks, AK: University of Alaska, Dec. 5, 1996.

Schwartz, C.C., S. D. Miller, and M.A. Haroldson. "Grizzly bear", Wild Mammals of North America: Biology, management, and conservation. 2nd ed. G.A. Feldhamer, B.C. Thompson, and J.A. Chapman, ed. Baltimore, MD: Johns Hopkins University Press, 2003.

Schindler, Daniel E. Mark D. Scheuerell et al. "Pacific salmon and the ecology of coastal ecosystems", Frontiers in ecology and the environment, Vol 1, No. 1, 33-37.

Scudder, H.C. The Alaska salmon trap: its evolution, conflicts, and consequences. Juneau, AK: Alaska State Library, 1970.

Shepard, Paul and Barry Sanders. The sacred paw. New York, NY: Viking, 1985.

Sherwonit, Bill. "Anan Creek Bear Observatory", Alaska Magazine, February, 1997.

Sherwonit, Bill, "Bear Viewing", Alaska's Bears. Anchorage, AK: Alaska Geographic Society. Vol. 20, No. 4, 1993.

The Silver Years. Lawrence Freeburn, ed. Anchorage, AK: Alaska Northwest Publishing Company, 1976.

Stewart, Hillary. Indian Fishing: early methods on the Northwest Coast. Vancouver, BC: Douglas & McIntyre, 1977.

Stowell, Harold H. Geology of Southeast Alaska. Fairbanks, AK: University of Alaska Press, 2006.

Swanton, John R. Tlingit myths and texts. Washington, DC: U.S. Government Printing Office 1909.

"Systematic looting of Point Ward Cannery", Wrangell Sentinel. January 23, 1913.

Thornton, Thomas F. Subsistence use of brown bear in Southeast Alaska: Technical paper number 214. Juneau, AK: Alaska Department of Fish and Game, 1992.

Tongass Land Management Plan Revision, Final supplemental environmental impact statement, Roadless area evaluation for wilderness recommendations, Vol. I, Appendices A, B, D,E. Washington, DC: United States Department of Agriculture, U.S. Forest Service R10-MB-481a, 2003.

Tongass Land Management Plan Revision, Final supplemental environmental impact statement, Roadless area evaluation for wilderness recommendations, Vol. II, Appendix C. Washington, DC: United States Department of Agriculture, U.S. Forest Service R10-MB-481b, 2003.

Trakowski, Bernice. "Fish Trap Patrol", The Alaska Sportsman, February 1953.

"UW biologist: spawning salmon behavior triggers stream ecosystem changes". Cyberwest Magazine, March 12, 2006. http://www.cyberwest.com/stream-ecology/alaska-salmon-spawning-behavior.shtml.

Vancouver, George. Voyage of discovery to the North Pacific Ocean and round the world., Vol. 2. London: G.G. and J. Robinson, 1798. [Reprint Amsterdam, Netherlands: N. Israel, 1967].

Waits, Lisette P., Sandra L. Talbot, R.S. Ward and G.F. Shields. "Mitochondrial DNA phylogeography of the North American brown bear and implications for conservation", Conservation Biology, Vol. 12, No. 2, April 1998.

Walker, Coose and Tammy Davis. "Dead salmon bring life to rivers", Alaska Wildlife News, Dec. 2004, http://www.wc.adfg.state.ak.us/index.cfm?adfg=wildlife_news.view_article&articles_id=97&issue_id=21.

Willson Mary F. and Karl C. Halupka. "Anadromous fish as keystone species in vertebrate communities", Conservation Biology, Vol 9, No. 3, June 1995, pp. 489-497.

Wing, Joel, "Southeast's First Cannery: 1867", Wrangell Sentinel, April 14, 1976.

Zadina, Timothy P.; Heinl, Steven C. et al. Pink salmon stock status and escapement goals in Southeast Alaska and Yakutat, Regional Information Report No. 1J03-06. Juneau, AK: Alaska Department of Fish and Game, Division of Commercial Fisheries, 2003.

# Index